Springer Textbooks in Earth Sciences, Geography and Environment

The Springer Textbooks series publishes a broad portfolio of textbooks on Earth Sciences, Geography and Environmental Science. Springer textbooks provide comprehensive introductions as well as in-depth knowledge for advanced studies. A clear, reader-friendly layout and features such as end-of-chapter summaries, work examples, exercises, and glossaries help the reader to access the subject. Springer textbooks are essential for students, researchers and applied scientists.

More information about this series at ►http://www.springer.com/series/15201

Martin H. Trauth
Elisabeth Sillmann

Collecting, Processing and Presenting Geoscientific Information

MATLAB® and Design Recipes for Earth Sciences

Second Edition

 Springer

Martin H. Trauth
Institute of Earth and Environmental
Science
University of Potsdam
Potsdam, Germany

Elisabeth Sillmann
BlaetterwaldDesign
Scientific Publications
Landau, Germany

Additional material to this book can be downloaded from ▶http://extras.
springer.com.

ISSN 2510-1307   ISSN 2510-1315   (electronic)
Springer Textbooks in Earth Sciences, Geography and Environment
ISBN 978-3-662-58572-6   ISBN 978-3-662-56203-1   (eBook)
https://doi.org/10.1007/978-3-662-56203-1

Printed on acid-free paper

This Springer imprint is published by the registered company Springer-Verlag GmbH,
DE part of Springer Nature
The registered company address is: Heidelberger Platz 3, 14197 Berlin, Germany

Preface

The book *Collecting, Processing and Presenting Geoscientific Information–2nd Edition*, known in its first edition as *MATLAB and Design Recipes for Earth Sciences–How to collect, process and present geoscientific information*, is designed to help undergraduate and postgraduate students, postdoctoral researchers, and professionals find quick solutions for common problems when starting out on a new research project. A project usually starts with searching and reviewing the relevant literature and data, and then extracting relevant information as text, data, or graphs from the literature, followed by searching, processing, and visualizing the data, and finally, compiling and presenting the results as posters, abstracts, and talks at conferences.

The course on which this book is based was first taught by M.H.T. as a bachelor's module for second-year students during the 2010/11 winter semester, three years after the introduction of bachelor's and master's programs at the University of Potsdam. The initial design of the bachelor's program included an introductory course on data analysis, scheduled for the second year, which was based on the sister book to this one: *MATLAB Recipes for Earth Sciences–4th Edition* (Trauth 2015). This course was a complete failure, probably because the second-year bachelor students were not well enough prepared for an advanced course on data analysis, even after two semesters of mathematics during the first year. A few weeks later, the course for students at master's and doctoral levels on the same topic, which M.H.T. was invited to give at the University of Ghent in Belgium, was a great success. The difference between the undergraduate students in Potsdam and the graduate students in Belgium was, of course, the greater motivation that students already working on their own projects had to learn the statistical and numerical methods offered by MATLAB, in order to be able to analyze their data.

As a consequence, M.H.T. moved the course into the master's program and designed a completely new course on *How to Collect, Process and Present Geoscientific Information*, which was very well received by the second-year students, despite the very large number of participants. The course was not presented as a complete package but evolved during the months of teaching, taking into consideration the suggestions made by students attending the course. During the course, and very much motivated by its success, the idea for this new book quickly emerged and the first outline for the text was drafted in late December 2010. Most of the text was written immediately following completion of the first course and before the start of the second course in spring 2011. Fortunately, the graphic design specialist E.S., who is M.H.T.'s sister and the owner of *blaetterwaldDesign*, joined the project to contribute to the design sections in the book as well as the book's layout, after having designed the layout of all four editions of the sister book *MATLAB Recipes for Earth Sci-

ences, as well as many other books, for *Springer*. The publisher quickly agreed to assist in realizing the book, and contracts were signed in summer 2011.

In the year 2016, we noticed that much of the book was outdated, the data sets used were replaced by more recent data sets, and many links to Internet sites were no longer valid. Furthermore, a new MATLAB® graphics system was introduced in 2014, which is based on an improved infrastructure, and although it supports most of the functionality of previous releases, there are some differences. We have therefore decided to publish an updated version of the book and also include some new material. We changed the title in order to distinguish it more clearly from the sister book. While the other book concentrates on data analysis, the MATLAB part of this book deals instead with the visualization of data, which is not included in the sister book.

While undergraduates participating in a course on data analysis might wish to work their way through the entire book, more experienced readers might refer to only one particular method in the book, in order to solve a specific problem. The concept of the book and its content are therefore outlined below, in order to make it easier for readers with a variety of different requirements to decide how they wish to approach the book.

►Chapter 1—This chapter is about initiating, planning, and organizing a project. It introduces the Internet resources used in the following chapters to search for geoscientific information, as well as the software and online tools used to manage projects, to process data, to exchange information, and to present project results.

►Chapters 2 and 3—These chapters deal with searching and reviewing scientific literature and data on the Internet. ►Chapter 2 provides a comprehensive tutorial-style introduction to Internet literature resources. It also demonstrates how to extract information from the literature for use within the reader's own projects and introduces software for managing large collections of electronic journal articles and books. ►Chapter 3 introduces the most popular data formats on the Internet, and methods to store and transfer such data. Data access and management is demonstrated by means of typical examples.

►Chapters 4–7—The first of these chapters starts with a tutorial-style introduction to MATLAB, designed for earth scientists (as in Chapter 2 of the sister book). ►Chapters 5 and 6 introduce advanced visualization techniques with MATLAB, for example how to create sophisticated two- and three-dimensional graphs from data collected in Chapter 3. ►Chapter 7 is on processing and displaying images with MATLAB, including satellite images (as in Chapter 8 of the sister book). The new edition contains sections about array manipulation, control flow, and the Live Editor introduced in 2016. The chapter on image processing now also includes a section on image enhancement, correction, and rectification.

►Chapter 8—The graphs created with MATLAB in the previous chapters are now handed over to the graphic design unit of the project. Even though the advanced plotting features of MATLAB presented in Chapters 5 and 6 are able to create sophisticated figures, all graphs will require further editing with vector and image processing software before they can be included, together with text and tables, in conference presentations and manuscripts.

►Chapters 9–12—These chapters are about creating effective conference presentations for talks or posters, and also various types of manuscripts for publication. They cover the preparation of colorful flyers and brochures relating to specific projects, as well as theses or project reports with relatively modest designs, and they also deal with assembling books and their layout design. Both Chapters 9 and 10 close with some remarks on practicing for conference presentations and their final delivery. ►Chapter 12 is a completely new chapter on multimedia publications.

The book contains *MATLAB* scripts or *M-files* for visualizing typical earth science data sets (►http://mathworks.com). The MATLAB recipes and data files for example used in the book (available online through Springer Extras, ►http://extras.springer.com) can be easily adapted to the reader's data and projects. M.H.T. developed these recipes using MATLAB Version 9 (release R2016b), but most of them will also work with earlier releases. The book also relies on numerous other software products, first and foremost being the *Adobe Creative Cloud* subscription service (►http://adobe.com), which is used to edit all the graphs created with MATLAB. The use of the Adobe Creative Cloud produces consistently high-quality results for all graphics to be included in project presentations. Nevertheless, it might be suitable to check and to transfer the proposed workflow to open-source alternatives as well, depending on ones individual needs. The book provides brief introductions to the use of Adobe graphics editors by means of step-by-step tutorials, supplemented by screenshots documenting the workflow that are provided online at Springer Extras.

We hope that our readers will appreciate our efforts to introduce open-source software tools in addition to the commercial products that the authors of this book use in their daily work. During the course at the University of Potsdam, students asked about free alternatives to *MATLAB*, such as *Python*, *R*, and *Octave*. Some students also liked to use *LaTeX* for typesetting, and *GMT* for creating *xy* and *xyz* plots. Students' financial resources are often limited, and many therefore use open-source software on their computers. For professionals, however, time is by far the more important limiting factor. When trying to meet a strict deadline for the submission of a research proposal or report, one quickly learns to appreciate complete and concise software manuals and the short response time of the software vendor's support line.

In putting together this book, we have benefited from the comments offered by many people, mostly students using the book. It is expected that this book will be constantly changing and evolving over time, as has been the case through the various editions of its sister book, *MATLAB Recipes for Earth Sciences*. As the Second Edition appears on the bookshelves, we will therefore create a new folder on the hard disk in preparation for the third edition. Please send us your comments and criticisms on the text, suggestions for correction and expansion of the text, and comments on any experiences that you may have had with similar courses or books. Please visit our Webpages ►http://martinhtrauth.de, ►http://mres.uni-potsdam.de and ►http://blaetterwalddesign.de from time to time, in order to check for updates and errata files for this book.

We are much obliged to Ed Manning for professional proofreading of the text. We would like to thank Annett Büttner at *Springer*, and also Andreas Bohlen, Brunhilde Schulz and their team at *UP Transfer GmbH* for their support. M.H.T. acknowledges the *Book Program* and the *Academic Support* at *The MathWorks Inc.*, including Sebastian Groß, Cedric Esmarch, and Andreas Himmeldorf. M.H.T. would also like to thank Mike Abrams and the *NASA/GSFC/METI/ERSDAC/JAROS* and the *US/Japan ASTER Science Team*, for permission to include their ASTER images in the book.

Martin H. Trauth
Geoscientist, University of Potsdam
Potsdam, Germany

Elisabeth Sillmann
Designer, blaetterwaldDesign.de
Landau, Germany

September 2017

Recommended Reading

Trauth MH (2015) MATLAB® Recipes for Earth Sciences–4th Edition. Springer, Berlin

Contents

About the Authors

Martin H. Trauth

was born in Landau in der Pfalz in 1963 and studied geophysics and geology at the University of Karlsruhe. He obtained a doctoral degree from the University of Kiel in 1995 and then became a permanent member of the scientific staff at the University of Potsdam. Following his habilitation in 2003, he became a lecturer, and then in 2011 an adjunct professor, at the University of Potsdam. Since 1990, he has worked on various aspects of historical changes in the climates of East Africa and South America. His projects have aimed to understand the role of the tropics in terminating ice ages, the relationship between climatic changes and human evolution, and the influence that climate anomalies had on mass movements in the Central Andes. Each of these projects has involved the use of MATLAB to apply numerical and statistical methods (such as time-series analysis and signal processing) to paleoclimate time series, lake-balance modeling, stochastic modeling of bioturbation, age-depth modeling of sedimentary sequences, or the processing of satellite and microscope images. Martin H. Trauth has been teaching a variety of courses on data analysis in earth sciences for more twenty years, both at the University of Potsdam and at other universities around the world.

Elisabeth Sillmann

has been providing design services for scientific media since 2004 with a focus on earth and environmental sciences. Her studio blaetterwaldDesign.de offers graphic design work, such as illustrations, books, presentations, conference posters, and Websites. Elisabeth Sillmann has extensive industrial experience in product development, product design, and project management. She works with international scientists and well-known publishers, providing a worldwide fast and efficient personal service, in English or German language.

Scientific Information in Earth Sciences

© Springer-Verlag GmbH Germany, part of Springer Nature 2018
M. H. Trauth and E. Sillmann, *Collecting, Processing and Presenting Geoscientific Information*, Springer Textbooks in Earth Sciences, Geography and Environment,
https://doi.org/10.1007/978-3-662-56203-1_1

1.1 **Introduction**

1

This book is based on an undergraduate course taught at the University of Potsdam in Germany (▶http://uni-potsdam.de), as was also the case with its sister book *MATLAB Recipes for Earth Sciences—4th Edition* (Trauth 2015). The objective of this course was to guide students through the typical progression of a scientific project. Such projects usually start with a search of the relevant literature in order to review and rank published books and journal articles, to extract relevant information (as text, data, maps, or graphs), and to search, process and visualize data, compiling the results and present them as posters, abstracts and oral presentations (talks). The course was first held for second-year students in earth sciences during the 2010–11 winter semester, and then repeated in the following summer semester. The original plan was to hold the course in a computer pool with fifteen workstations. However, an unexpectedly large number of students enrolled for the first presentation, which had more than sixty participants. This led to the course being held in a lecture hall with a projector, a microphone, and a speaker system. There was also a table for the instructor's laptop and equipment, and wireless access to the Internet; the students used their own private laptops.

The change of teaching rooms had both pros and cons. The large lecture hall of course provided a nice conference-type atmosphere with its audio-video system, light dimming and room darkening systems, and large projection screens. The use of private laptops had the advantage that the students were already familiar with their own computing systems prior to the start of the course. At the end of the course, the participants could carry home their entire workspace and continue the course-related project work at home. On the other hand, a clear disadvantage was that the students' private laptops had hardly any of the required software packages installed, such as *MATLAB*, the *Adobe Creative Cloud* and *Microsoft Office*. The obvious solution to this problem was the consistent use of *Open Source Software* instead of commercial products, accepting the clear limitations of most of these free software tools such as the often limited range of functions and performance, incomplete documentation and the general lack of support.

The compromise made in the course was to use *MATLAB* commercial software, which is also widely-used in many other courses and is becoming increasingly popular in earth sciences, rather than *Octave*, *Python* or other free products. The *Adobe Creative Cloud* (including *InDesign* for desktop publishing, *Illustrator* for editing vector graphics and *Photoshop* for editing pixel graphics), was, however, largely replaced by the free *Scribus*, *Inkscape*, and *Gimp* software. As a medium-priced alternative, at least for Apple's macOS and Microsoft Windows operating systems, the award-winning *Affinity* suite by Serif Ltd., including *Affinity Designer* (for editing vector graphics), *Affinity Photo* (for editing pixels graphics) and *Affinity Publisher* (for desktop publishing), was recommended to the students. *Microsoft Office* (including *Word* for text processing, *Excel* for spreadsheet calculations and *PowerPoint* for presentations), and the Apple equivalent, *iWork* (including *Pages*, *Numbers* and *Keynote*), were replaced by the free *OpenOffice* suite. Most students purchased the inexpensive *MATLAB Student Version* for their project work, as well as downloading and installing the recommended free software packages. In this book, however, not being a software-oriented tutorial, we use both commercial and free products to demonstrate the principles for collecting, processing, and presenting information in earth sciences.

A major challenge was to handle the large number of participants in a highly interactive, tutorial-style environment. The solution was to establish a reliable eLearning environment for the course using the free *Modular Object-Oriented Dynamic Learning Environment* (Moodle, ▶http://moodle.org), also known as a *Course Management System* (CMS) and widely-used for electronically-supported learning and teaching. The Moodle course page provided a section for each weekly class including presentations, teaching manuscripts and additional course materials such as literature, data, example files, and interesting data and information links. The course page also included a message board for announcements, a forum for discussions and drop boxes for homework. We used *Etherpad* (▶http://etherpad.org), or the alternative *Pad.UP*, at the University of Potsdam for online editing of text, such as MATLAB code and students' texts written during the course.

In addition to establishment of the Moodle and Etherpad course environments, the students were organized into three groups using *Google Groups* (▶http://groups.google. com), *Dropbox* (▶https://www.dropbox.com), or an alternative file hosting service such as *Box.UP*, provided by the University of Potsdam. During the course, the instructor defined three principal topics for each of the groups, within the overall theme of *Past Climate Changes*. One advantage of having the students organized in groups was the introduction of teamwork; collecting literature and data, processing the results, solving minor problems (e.g., technical problems), and organizing the presentations at the end of the course all worked exceptionally well within the groups. The instructor was mainly contacted by the group leaders to answer more fundamental questions and to provide support with organizing and managing the teamwork within the groups.

The following information on the eleven-week course is provided as it may be useful for instructors intending to design their own course on the basis of this book. There was a three hour class in each of these weeks with lectures, demonstrations, student presentations, and discussions. The typical work load for each student was 8–10 h per week, including the three-hour class; this resulted in 6 credit points under the European Credit Transfer System. The typical syllabus for the courses on *Collecting, Processing and Presenting Geoscientific Information* was as follows:

- Week 1—Introduction and class on ▶Chap. 1: *Scientific Information in Earth Sciences*. Explanation of the course concept, description of the course examination procedure (including the submission of a 200-word abstract, the design of a poster, and the delivery of a two-minute presentation in the last three weeks of the course), definition of the group's research topics, establishment of groups and election of the group leaders.
- Week 2—Presentations by group leaders on the establishment and organization of each group (approximately 5 min each). Highly interactive class on ▶Chap. 2: *Searching and Reviewing Scientific Literature*.
- Week 3—Presentations by group leaders of results from their searches for the five most relevant articles on their research topics. Class on ▶Chap. 3: *Internet Resources for Earth Science Data*.
- Week 4—Reports by group leaders or their deputies (5–10 min each) on the teamwork within the groups and how they organize and rank the information collected

in the groups. Presentation of the 10–20 most important articles on their research topics, including a description of the system used to rank these articles and an attempt to display the list in a structured way. Class on ▶Chap. 4: *MATLAB as a Visualization Tool*.

━ Weeks 5–7—Reports by each group (one per week, 10–15 min each) on the most exciting hypotheses and controversies in their research topics, presenting the relevant references, representative graphs, and data sets. Presentations of summaries by each group, with the presentation method being chosen by the group and with the presentations being made by either a single representative (the group leader, a deputy, or any other group member) or, as some of the groups preferred, by a group of three to five representatives, each presenting different parts of the summary. At this stage of the course the groups start to split their research topic into sub-themes, each of which is pursued by an individual member of the group. Classes on ▶Chaps. 5–7: *Visualizing Data and Processing and Displaying Images in Earth Sciences*.

━ Week 8—Presentation by each group of an outline for the conference session to be held at the end of the course, followed by discussion of the sub-themes and the proposed titles of the talks. Class on ▶Chap. 8, *Editing Text and Tables*, ▶Chap. 11, *Creating Manuscripts*, and ▶Chap. 12, *Creating Multimedia Publications*, since the first student work to be submitted will be the conference abstract.

━ Week 9—Presentation by each group of an updated outline of the conference session. The groups may also present drafts of individual graphs to be included in posters and presentations. Class on ▶Chap. 8, *Editing Vector and Raster Graphics* and ▶Chap. 10, *Creating Conference Posters*. Student examination: each student posts to the Moodle platform a 200-word abstract on their sub-theme.

━ Weeks 10—Presentation by each group of an updated outline of the conference session, and discussion of the schedule for the conference session. The groups can also present drafts of individual conference slides. Class on ▶Chap. 8, *Editing Vector and Raster Graphics* and ▶Chap. 9, *Creating Conference Presentations*, and ▶Chap. 12, *Creating Multimedia Publications*. Student examination: each student presents a poster on their sub-theme, including the previously submitted abstract.

━ Week 11—Conference sessions organized by the students. Student examination: each student gives a two-minute presentation in a large lecture hall, in front of a jury of three professors.

Each of these components have combined to form the current design of the course and the concept behind this book. At the University of Potsdam the course culminated in the presentation of the project work as conference-style posters, 200-word abstracts, and one-hour sessions with 10–15 two-minute presentations, chaired by the project leaders and their deputies. The following sections of this chapter provide a closer look at the design of the course and this companion book, introducing briefly the techniques for collecting, processing and presenting information in earth sciences, as well as the information formats encountered while working with data in our discipline and strategies for optimizing and organizing the workspace and work flow.

1.2 Collecting and Managing Information in Earth Sciences

1.2.1 Initiating, Planning and Organizing a Project

Imagine a small group of people (perhaps two or three) with an idea for a new research project. Having identified a specific topic as a research area worthy of investigation, the initial group starts putting together a team of researchers with the necessary methodological expertise to run the project. After assembling the team, the core group behind the new project writes a research proposal for submission to a funding agency. The proposal is reviewed and, if successful, funding is provided, after which the project can actually start.

During the course we simulated the initiation, planning, and organization of a research project. In the summer semester course the instructor defined three research themes as examples of the topics for student projects:

- "Tectonics and climate",
- "The ice ages during the Pleistocene epoch", and
- "Climate through the last millennium".

The instructor then asked the students to designate a team leader for each of the three topics. The students accepted as team leaders then established Google Groups. According to their webpage, Google Groups was started in 2001 and helps project members connect with other members, access information, and communicate effectively by email and on the web. The service provides a mailing list for each group to use for project discussions. Google Groups is linked with *Google Docs* (►https://www.google.com/intl/en/docs), which became available in 2007 and in which you can create and share text documents, spreadsheets, presentations, and many other types of files so that they can be accessed from anywhere. As an alternative to Google Docs, the students also used *DropBox* (►http://dropbox.com), which is a free web-based file-hosting service. DropBox uses a cloud-storage system to synchronize files on different computers with a server via the Internet.

After establishing the groups, the team leaders were then contacted by the remaining students requesting participation in the groups. Each group then selected deputy team leaders and soon after that started forming sub-groups to work on specific aspects or sub-themes of the overall research themes. As an example, the group working on *Tectonics and Climate* as a research topic defined sub-themes such as

- "Mountain uplift and climate change",
- "Closure and opening of ocean gateways and climate change", and
- "Volcanic activity altering the Earth's climate".

The sub-themes were identified by the students themselves soon after they started searching and reviewing the literature on the overall research topic of their group.

1.2.2 Literature and Data Resources on the Internet

The Internet provides fast access to geoscientific information such as conference proceedings, electronic books and journal articles, digital geologic maps, satellite images,

1

and earthquake catalogs. ►Chapters 2 and 3 demonstrate the use of literature search catalogs and various other resources for accessing earth science data.

In 2011 the online book store *Amazon* (►http://amazon.com) announced that they had sold more electronic books than printed books. Long before that most researchers and their institutions had switched from printed to electronic collections of journal articles. The advantage of an electronic collection is obvious as it allows large volumes of journal articles to be easily transported while traveling. Data format incompatibilities between different computer operating systems or software types are fortunately largely resolved by cross-platform formats such as the *Portable Document Format* (PDF), introduced in 1993 by Adobe Systems, for electronic books, reports and journal articles. Similarly, the standard formats for digital images such as the TIFF, PNG or JPEG allow images to be exchanged between computers via the Internet or storage devices, and the establishment of large image galleries on the Internet (►Chap. 7).

The amount of geoscientific information available on the Internet is enormous. Web search engines such as the market leader *Google Search* (►http://google.com) or the less popular *Yahoo!* (►http://yahoo.com) and Microsoft's *Bing* (►http://bing.com) rank webpages by their relevance using patented algorithms and can therefore respond quite efficiently to queries by Internet users. More specific catalogs exist for searching, evaluating and accessing literature such as the free *Google Books* (►http://books.google.com) and *Google Scholar* (►http://scholar.google.com), and commercial products such as Elsevier's *Scopus* (►http://scopus.com) or Thomson Reuters *Web of Science* (►http://apps.webofknowledge.com) (►Chap. 2). Geoscientific data are available from large data servers such as the US *National Centers for Environmental Information* (NCEI), formerly known as the US *National Climate Data Center* (NCDC) and the US *National Geophysical Data Center* (NGDC) (►http://ncdc.noaa.gov) or the German *Pangaea Data Publisher for Earth & Environmental Science* (►http://pangaea.de) (►Chap. 3).

1.2.3 Project Management

In this section we introduce some useful tools for organizing projects with many participants, for controlling the workflow of the project, and for communicating online within the project. We also demonstrate the use of tools to manage the geoscientific information that is collected during the course of the project.

Google Groups, in combination with Google Docs, provides a useful platform for communication within a project, for collecting and ranking information, and for processing documents such as project reports, books, or journal articles. During the last few years *Facebook* (►http://facebook.com, launched by Dustin Moskovitz, Chris Hughes, Eduardo Saverin und Mark Zuckerberg in 2004) has also become increasingly popular for presenting research projects on the web. Similar to Facebook but specifically for scientists, *ResearchGate* (►http://researchgate.net, launched in May 2008 by physicians Ijad Madisch and Sören Hofmayer, and computer scientist Horst Fickenscher) provides a social-network for used by than 10 million scientists (October 2016) to share papers, ask questions and find possible collaborators. In addition to these popular platforms, numerous other networks exist that are mostly used by specific disciplines rather than being widely used Internet tools for project communication in general.

To coordinate project meetings, *Doodle* (▶http://doodle.com) has become very popular since it was launched by Michael Näf in 2003. The project leader instigates a Doodle poll by suggesting possible time slots for the meeting and inviting all project members to indicate those dates on which they are available. The project leader then closes the poll and specifies the date selected for the meeting on the basis of mutual availability. Time management during the course of the project is also of great importance and can often be a source of conflict. It is important for the project leader to adhere strictly to the schedule in order to avoid the frustration of individual project members that rely on each other's results, and to avoid any extra costs due to delays. Numerous free calendar and time management software tools, partly cloud-based, are available, as well as commercial alternatives, but will not be discussed here in further detail.

The financial management of a project is typically the responsibility of administrative staff at universities and research institutes. However, the project leader also needs to keep track of the funds that have been spent because of the inherent delay between ordering materials and payment of invoices. Most of the spreadsheet software listed in ▶Sect. 1.3 can be used for managing finances but more advanced accounting software (not discussed here) may also be useful for managing larger projects.

During the course of the project all kinds of data accumulate (both original and processed) and also need to be managed. Data management in larger projects is a challenging task that requires the use of database software tools rather than a simple system of files and folders on a hard drive. In this context, a reliable backup system also needs to be established comprising backup hard disks, such as a *Redundant Array of Independent Disks* (RAID) that tolerates the complete failure of one out of several disks, and backup software such as the Apple's *Time Machine* software. Storing a backup hard disk in a separate place from the place where the data are processed is advisable. Collaborating on data and documents requires the strict use of version numbers (see also ▶Sect. 3.9) or use of the online tools provided by Google Docs, which allow documents to be edited concurrently by multiple users.

1.3 Methods for Processing Scientific Information

Geoscientific information, whether generated within the project or derived from other resources, can be digitally processed by many different kinds of computer software tools. These tools include software for transferring the actual information from one computer to another, from a storage device to computers, or between computers and remote servers in a larger network. The data are processed either locally on a desktop or laptop computer, or remotely on a server using a text terminal or console, a *graphical user interface* (GUI) based tool, or web-based tool. The processed data are then included in posters, presentations, or papers and other publications.

1.3.1 Software for Transferring Scientific Information

Since the establishment of computer networks, ever increasing quantities of data are transferred from computer to computer via the Internet. Before the introduction of the

World Wide Web (WWW) and the first web browsers, digital information was mainly accessed through a network by the entry of commands via a keyboard in a text terminal or console (see also ►Sect. 3.4). Examples of text terminal applications are *Terminal* and *Console* for UNIX-based operating systems such as Solaris, Linux, or Apple macOS, and *cmd.exe* or *PowerShell* for Microsoft Windows.

Whereas in the early 1990s we used these text terminals to send emails and to download data, most scientists today use GUI-based tools such as email clients, FTP software, and web browsers to access and share digital information over the Internet. Popular email clients include the open source cross-platform Mozilla *Thunderbird* (►https://wiki.mozilla.org/Main_Page), the Microsoft Windows *Live Mail* which is the successor to the popular *Outlook Express* and Windows *Mail* clients (►https://www.microsoft.com/windows), and the Apple *Mail* included in Apple's macOS operating system (►http://www.apple.com/de/macos).

Data transfer using the *File Transfer Protocol* (FTP) is usually achieved through GUI-based programs such as the popular *Cyberduck* available for Windows and Mac (►https://cyberduck.io), the free *FileZilla* (►http://filezilla-project.org), or the commercial *WS_FTP* (►http://ipswitchft.com), but numerous other free and commercial alternatives also exist (see also ►Sect. 3.4). In many cases, however, the user is not aware of the change in protocols while browsing the Internet to search for geoscientific information online. Since the introduction of the first web browser, *NSCA Mosaic*, in 1993, which later became the *Mozilla* suite of web tools, the most popular web browsers, based on the list of market shares by *NetMarketShare* (September 2016, ►https://netmarketshare.com) are the cross-platform *Chrome* by Google (►http://google.com/chrome) with more than half of the market share, the Microsoft *Internet Explorer*, recently replaced by *Edge* (►https://www.microsoft.com/windows) included in Microsoft Windows (but discontinued for Mac and UNIX), the cross-platform *Firefox* (►http://mozilla.org), and *Safari* by Apple macOS (►http://apple.com/safari).

1.3.2 Software for Processing Scientific Information

By far the most popular software for processing data is the *Microsoft Excel* spreadsheet software (►http://office.microsoft.com/excel). Following its first introduction for Macs in the mid-1980s, and later for Microsoft Windows, the software has seen multiple modifications and improvements before being included in the *Microsoft Office* suite, which also contains *Microsoft Word* and *Microsoft PowerPoint* (►http://office.microsoft.com). In 2007 Apple introduced *Numbers* as an alternative to Excel and included it in the Apple *iWork* suite (►http://apple.com/iwork). Numbers is compatible with Excel, but more stable and faster on Macs. A free open-source alternative to Excel is *Open Office Calc*, which is included in the cross-platform *OpenOffice* suite that also contains *Writer* and *Impress* (►http://openoffice.org). The more advanced, but also more expensive alternative to Excel is the popular *SPSS* software package, first released in 1968 and acquired by IBM in 2009 (►http://www.ibm.com/analytics/us/en/technology/spss).

Spreadsheet software packages are, however, generally very limited in their ability to process large data sets, such as those for digital elevation models or images. Various high-level programming languages integrated into software environments, either

GUI-based or solely command-line based (or both), are very popular tools for processing scientific information contained in such data sets. The programming language *R*, which was first introduced in 1993, is probably the most popular member of this family of tools (►http://r-project.org) and is also free. R has a command-line interface although several GUIs are available; it has a large number of available libraries and very good documentation. The cross-platform *Python* (►http://python.org), first introduced in 1991, is also free and command-line based, but with graphics output using *Matplotlib Graphics* (►http://matplotlib.sourceforge.net). The *Interactive Data Language* (IDL) is particularly popular in physics, including astronomy and medical imaging; it was introduced in 1977 and is now sold by Harris Geospatial Solutions (►http://www. harrisgeospatial.com).

According to the webpage of The MathWorks Inc., the commercial *MATLAB* software is a high-level technical computing language and interactive environment for algorithm development, data visualization, data analysis, and numeric computation (►http://mathworks.com). The original software goes back to the late 1970s when Cleve Moler, co-founder of The MathWorks Inc., wrote a program for performing numerical linear algebra. MATLAB can be run either command-line based or using a graphical user interface (GUI). Since the cross-platform MATLAB software contains a large library of ready-to-use routines for a wide range of applications, the user can solve technical computing problems much more quickly than with traditional programming languages such as C++ and FORTRAN. The standard library of functions can also be significantly expanded by add-on toolboxes, which are collections of functions for special purposes such as image processing, creating map displays, performing geospatial data analysis, or solving partial differential equations. MATLAB is relatively inexpensive for students, either as the MATLAB Student Version (►https://www.mathworks. com/academia/student_version) which can be ordered by individual students, or as a Student Group License for multiple users, which can be ordered by course instructors thus further reducing the cost per license. The cross-platform *Octave* language is very similar to MATLAB and most routines can be easily ported between the two (►http:// www.gnu.org/software/octave).

MATLAB was primarily designed for numerical computations. The *Symbolic Math Toolbox* introduced in 1993 provides tools for solving and manipulating symbolic math expressions and performing variable-precision arithmetic. The classic alternative for such tasks has always been the cross-platform commercial *Mathematica* software developed in 1988 by Stephen Wolfram, who was also the co-founder of Wolfram Research (►http://wolfram.com). Over the last two decades, however, Mathematica has also become a powerful tool for high-performance numerical computing similar to MATLAB but with a completely different syntax.

This book does not deal directly with geographical information systems (GIS) or remote sensing. We will, however, use and process digital elevation models and create maps in ►Chap. 6 using MATLAB. We will also process and georeference a satellite image in ►Chap. 7, again using MATLAB. More advanced applications in GIS or remote sensing may, however, require software specifically designed for those particular applications. The market leader for GIS is the software development company ESRI (►http://esri.com) that makes the commercial ESRI product suite, which includes *ArcGIS*. Free alternatives for GIS applications are *GRASS* (►http://grass.osgeo.org) and

QGIS, formerly known as *Quantum-GIS* (►http://www.qgis.org). The most popular types of commercial remote sensing software are *ERDAS IMAGINE* (►http://www.hexagongeospatial.com) and ITT Visual Information Solutions *ENVI* (►http://www.harrisgeospatial.com).

I hope this brief overview of the most popular software packages will help the reader to choose the most suitable software for his/her applications. My personal preference is for MATLAB because this software makes it relatively easy to get started with sophisticated tools of numerical computing and to create attractive and publishable graphics. I cannot say much about Python but I understand from my colleagues in physics that it is also very popular in certain fields. Octave is certainly an attractive free alternative to MATLAB but has the disadvantage that it lacks the excellent support provided by The MathWorks Inc. Coming from FORTRAN, I did not enjoy the cryptic syntax of Mathematica but nevertheless used it until the mid 1990s, for example to invert filters using a Taylor series before then using MATLAB for the actual filtering.

1.3.3 Software for Editing Raster/Pixel Graphics

The *Photoshop* software, part of the *Adobe Creative Cloud* subscription service, is probably the most widely used professional software for editing raster/pixel graphics, mostly photos (►http://adobe.com/photoshop). Originally developed by Thomas Knoll, the software was purchased by Adobe in the late 1980s and first released for Macs in 1990. This software provides an industry-standard editing, enhancement, and output tool for photos. Some of the highlights of the software include excellent selection capabilities, content-aware fill options, processing tools for raw images, automated lens corrections, free-form transformations, advanced painting technologies, and the ability to create 3D artwork from selected graphics objects with direct control over lighting, materials and meshes.

Compared to this huge software package, the very popular and inexpensive *GraphicConverter* by *Lemkesoft GmbH* is a relatively spartan and light product (►http://lemkesoft.de). According to the company's webpage, the founder of the company, Thorsten Lemke, changed from Atari to Macintosh in 1992 and wanted to convert his picture collection to a Mac format. Exclusively available for Macs, the software has 1.5 million users worldwide. As a medium-priced alternative, at least for Apple's macOS and Microsoft Windows operating systems, the award-winning *Affinity* suite by Serif Ltd. includes *Affinity Photo* (released in 2015) for editing pixel graphics (►https://affinity.serif.com/en-us/photo).

In contrast to these commercial products, *GIMP* is a free multi-platform image processing software whose graphical user interface looks similar to that of Photoshop (►http://gimp.org). GIMP stands for GNU Image Manipulation Program, which can be used for painting, photo retouching, batch processing, and as an image format converter. The original version of the software was developed by Peter Mattis and Spencer Kimball and it was first released in 1996. Since Photoshop and GraphicConverter do not exist for Linux, GIMP is the most popular software on computers running under this operating system despite being much slower and less convenient when working with photos.

1.3.4 **Software for Editing Vector Graphics**

The *Illustrator* software, which is also part of the *Adobe Creative Cloud* subscription service, is the most advanced (but expensive) software for editing vector graphics (▶http://adobe.com/illustrator). The software was developed by Adobe in 1986 as a commercial version of the company's font development software using the PostScript file format (see ▶Sect. 3.3) developed by John Warnock at PARC (the Xerox research institute) with support from Aldus (now a part of Adobe Systems) and Apple Computers (now Apple Inc.). Apart from all the standard vector graphics tools, the software allows perspective drawing, variable-width strokes, stretch controls for brushes, bristle brush, crisp graphics for web and mobile devices, and many other advanced tools for vector graphics. As a medium-priced alternative, at least for Apple's macOS and Microsoft Windows operating systems, the award-winning *Affinity* suite by Serif Ltd. includes *Affinity Designer* (released in 2014) for editing vector graphics (▶https://affinity.serif.com/en-us/designer).

In contrast to Illustrator and other commercial products, the cross-platform *Inkscape* software is free (▶http://inkscape.org). This open source software was released in 2003 by Ted Gould, Bryce Harrington, Nathan Hurst and MenTaLguY, all members of the former SodiPodi project, which was also an open source vector graphics editor, discontinued in 2004. According to the software webpage it has capabilities similar to those of Illustrator, using the W3C standard Scalable Vector Graphics (SVG) file format. Having used Inkscape when teaching the course on data processing, I personally found that, while it had most of the functionality of the commercial Illustrator, it was far less stable and much slower, as well as being significantly less intuitive and practical. Inkscape, however, seems to be the state-of-the-art vector graphics software if you are using Linux as the operating system of your computer.

1.3.5 **Software for Creating Presentations**

The *PowerPoint* software, part of the *Microsoft Office* suite, is the market leader in presentation software and is available for both Microsoft Windows and macOS (▶http://office.microsoft.com/PowerPoint). It was created by Thomas Rudkin and Dennis Austin at Forethought Inc. and first released for Macs under the name Presenter. Renamed as PowerPoint, it was purchased by Microsoft in 1987 and later included in the Microsoft Office suite. The term *PowerPoint presentation* is now used for any general presentation consisting of a number of slides presented by a projector, even though numerous alternatives, both commercial and free, are now also available. The slides may typically contain a mixture of text, tables, graphics, sound, or videos. The objects within a slide can be arranged using a simple vector graphics editor. This easy-to-use vector graphics editor has made PowerPoint a popular and relatively inexpensive way to create graphics, and even posters, if Illustrator and similar products are not available. Critics of the software argue that the introduction of bullet point lists with PowerPoint encourages less conscientious presenters to oversimplify complex facts into a short list of statements and to rely on recurring, often boring, template designs included in the software package for their presentations (see *Death by PowerPoint*, a term first used by Angela R. Garber at Small Business Computing, ►http://smallbusi-nesscomputing.com).

As far as better template designs are concerned, the *Keynote* software included in the Apple *iWork* suite (►http://apple.com/iwork) aims to provide more attractive pres-entation templates (called themes) to the software user. Moreover, Keynote includes a much larger collection of animations than does PowerPoint, including animated tran-sitions between slides, although in science presentations these animations should be kept to a minimum and used only when appropriate for improving the presentation of scientific results. Keynote can import and export PowerPoint presentations and many Mac users therefore work with Keynote and then convert their presentations for con-ference laptops.

Impress, which is included in the OpenOffice suite, is a free alternative to these commercial products. Formerly known as StarOffice, OpenOffice was developed by StarDivision in the mid-1980s but later acquired by Sun Microsystems in 2002 and then by Oracle in 2010; it is now owned by the Apache Software Foundation. Impress essentially copies PowerPoint, with full compatibility. Since it was available for UNIX-based systems at a very early stage, the software is very popular for LINUX computers.

1.3.6 Software for Text Processing

Text processing is certainly one of the oldest applications for computers and it is there-fore not surprising that a large number of software tools are available for editing text. The most popular text processor is *Word*, as part of the *Microsoft Office* suite (►http://office.microsoft.com/word). The first version was created by Charles Simonyi in the early 1980s for Microsoft DOS; and subsequently released for Macs in 1985 and for Microsoft Windows in 1987. In 2005 Apple introduced *Pages* as an alternative to Word and this was later included in the Apple *iWork* suite (►http://apple.com/iwork). Pages is compatible with Word but more stable and faster on Macs. The *OpenOffice Writer* software included in the OpenOffice suite is a free alternative to these commercial products. Low-level text processors included in the respective operating systems are the Apple *Text Edit* and Microsoft *WordPad*, and there is also *Text Wrangler* (►http://barebones.com/products/textwrangler) or *XEmacs* (►http://xemacs.org), which are free text editors for use with programing tools. *MathType* by *Design Science* (►http://dessci.com) is a software for creating mathematical notations for word processors.

1.3.7 Software for Desktop Publishing

Desktop publishing describes *WYSIWYG* (an acronym for *what you see is what you get*) types of document creation for publication including text, tables, graphics and images. Even though most advanced text processors, such as Microsoft Word and Apple Pages, enable the user to create complex documents that include various types of text and graphics objects, large scale publishing of books, magazines, theses and other multi-page manuscripts requires the use of even more advanced software. In the mid-1980s Macs were the state-of-the-art desktop publishing instruments, especially following the introduction of the *PostScript* standard, the first Apple *LaserWriter*, and software products such as the Aldus *PageMaker*. Today, the current successor to Page-Maker is *InDesign*, which is part of the *Adobe Creative Cloud* subscription service and offers the most advanced (but expensive) software for desktop publishing (►http://adobe.com/indesign). This book has been created by the designer and co-author, Elisabeth Sillmann, using this software.

As a medium-priced alternative, at least for Apple's macOS and Microsoft Windows operating systems, the award-winning *Affinity* suite by Serif Ltd. includes *Affinity Publisher* (released in 2017) for desktop publishing (►https://affinity.serif.com/en-gb). *Scribus* is a free desktop publishing software that is similar to InDesign but is not as convenient, stable, or fast as the commercial product (►http://scribus.net). Scribus, however, is also available for Linux-based computers and therefore very popular in the Linux community.

1.3.8 Software for Managing Electronic Libraries on Computers

In order to organize a collection of references and electronic documents, we need a literature or bibliographic management software. A large number of open source and commercial software products are available for this purpose. The most popular commercial software products for literature management are Clarivate Analytics *Endnote* (►http://endnote.com) and Labtiva's *Papers* (►http://papersapp.com). Alternatively, software based on the *BibTex* standard such as *BibDesk* (►http://bibdesk.sourceforge.net) for the Unix-based macOS or *KBibTex* for Linux or Unix uses *LaTeX* to prepare formatted literature lists (►http://home.gna.org/kbibtex).

1.4 Presenting Geoscientific Information

The results of a project are typically presented in three formats: on posters, as talks, and in papers. In addition to these principal formats, various other more specific formats exist such as reports, theses, and books as additional examples of scientific information in a text format, keynote lectures as examples of more extended oral presentations, and web presentations including audio or video available online.

Young scientists at the beginning of their scientific careers usually present the first results of their research project as a *poster* at a workshop or conference. A poster is collection of graphs, photos, and text printed on a large piece of paper that is presented on a poster board, usually in a large hall within a conference building. During poster sessions the presenter of the poster has the opportunity to personally interact with people attending the poster session and visiting the poster board. Prior to the actual presentation of a poster, the scientist submits an *abstract* summarizing the key findings of the research project to date. Abstracts are typically limited to between 100 and 200 words, although exceptions are sometimes made (for example with extended abstracts, which can in some cases reach the length of journal articles). These brief summaries are included in conference volumes or catalogs in which all posters and oral presentations are listed. Abstract volumes are these days no longer printed, but the abstracts are provided on memory sticks or compact disks. Abstracts are viewed by conference session conveners or organizers and checked for relevance to the theme of the session and for their overall quality. In contrast to journal articles, however, abstracts are not peer reviewed. ►Chapter 10 demonstrates the design of a conference poster and ►Chap. 11 explains how to write an abstract.

Abstracts are also required to be submitted for an *oral presentation* or *talk*. Such an oral presentation typically lasts 15–20 min and is supported by a series of pages or slides projected using a video projector. The slides contain graphs and photos, but a relatively small amount of text. The talks are organized in theme sessions that are chaired by a session convener. The convener introduces the presenter, takes care of the time management for the session, and moderates the discussion with questions and answers after the presentation. ►Chapter 9 explains the planning of a presentation and the design of slides, as well as making suggestions for practicing and delivering the presentation.

After presenting the research results at conferences, either as a poster or as a talk, the research is published as a *paper*, report, thesis, or book. A paper is a journal article that is usually peer reviewed by colleagues who are experts in the same field of research but are not associated with the particular project. Papers are written text, generally four to ten printed pages long including graphs, photos and tables. They are submitted as Word or PDF files with figures, photos and tables included at the end of the file or attached as a separate file, mostly via the web interface of the journal's website. The manuscript arrives in the journal's editorial office where the editor first views and evaluates the paper for its overall quality, length, and suitability for the particular journal. If the paper passes this process it is sent to two or three independent reviewers from the same field of research who are qualified to evaluate the work. The reviewers, who in most cases remain anonymous, are requested to complete their review within two to four weeks and submit a recommendation to accept the manuscript as it is, to resubmit the manuscript with minor, moderate, or major revisions, or to reject the manuscript either with or without the possibility of resubmission after performing additional experiments and/or a complete rewrite of the text. The editor then collects the reviews, writes an editorial letter summarizing the assessments by the reviewers, and decides on the acceptance or rejection of the work. If it is accepted, the manuscript is then transferred to the copy editor's office where it is prepared for publication. ►Chapter 11 explains how papers are structured, written and prepared for submission to an international journal for review. ►Chapter 12 is about creating multimedia publications, which are becoming increasingly popular with, for example, the improving availability of electronic books (ebooks). Scientific journals such as Nature and Science also publish articles that include interactive objects (e.g. interactive rotatable molecules) within PDF files.

The author of an abstract or paper includes the published work in his/her list of publications, which is a record of the researcher's scientific output. The list of publications is included in the curriculum vitae of a scientist and, together with a list of funded research projects and a record of teaching experience, is an important prerequisite when applying for a research position at a research institute or university.

Recommended Reading

Trauth MH (2015) MATLAB® Recipes for Earth Sciences–4th Edition. Springer, Berlin

Searching and Reviewing Scientific Literature

© Springer-Verlag GmbH Germany, part of Springer Nature 2018
M. H. Trauth and E. Sillmann, *Collecting, Processing and Presenting Geoscientific
Information*, Springer Textbooks in Earth Sciences, Geography and Environment,
https://doi.org/10.1007/978-3-662-56203-1_2

2.1 Introduction

This chapter is on searching relevant literature, reviewing and ranking published books and journal articles, and extracting relevant information in the form of text, data, or graphs. In this context, the focus of our book is on Internet resources and literature in an electronic format such as the Adobe *Portable Document Format* (PDF), rather than on printed journals and books in a library. Although some of you might have a printed version of this book in your hands, most people have probably taken advantage of Springer's eBook Collection to read the digital version on a computer.

The advantages of electronic resources are very obvious, as drawers of suspension files filled with hundreds or thousands of printouts of journal articles, or office shelves containing large numbers of heavy books are of little use when working at home or traveling to conferences. This chapter focuses on the most popular open source and commercial Internet resources. The reference search described in the following section demonstrates a typical literature survey by students, starting with the popular online encyclopedias and bookstores, continuing to the generic web search engines to find relevant literature, and ending up in commercial literature data bases designed for professional use.

2.2 Resources for Literature Reviews

In this section, we go through a typical information and literature search, starting with the very popular web-based, open source encyclopedia *Wikipedia* (▶http://www.wikipedia.org) to find information on earth science topics. Wikipedia was launched in 2001 by Jimmy Wales and Larry Sanger. It consists of more than 40 million articles in more than 250 different languages and as of February 2014, it had 18 billion page views and nearly 500 million unique visitors each month, according its own article about the site. It is written and edited by largely anonymous Internet volunteers, under the supervision of experienced editors who ensure that any edits are cumulative improvements. Although often criticized for most articles not being edited by experts in the particular topic, the obvious advantage of Wikipedia is the topicality of its content.

Searching for literature, especially when trying to familiarize oneself with a completely new topic, usually starts in a bookstore. We have agreed to use Internet resources in this book and hence the bookstore will be online. The largest online retailer for books, and more recently for many other products including music, consumer electronics, and even food, is the online commerce company *Amazon* (▶http://www.amazon.com). Amazon was founded in 1994 by Jeff Bezos. In 2007 Amazon launched the Amazon Kindle, an ebook reader that can download content from the web via a wireless network.

Google was set up in 1998 by Larry Page and Sergey Brin as a graduate student research project at Stanford University. In addition to the general web search services, Google also hosts *Google Books* (▶http://books.google.com), which was introduced in 2004 under its original name of Google Print. This service searches the full text of both original ebooks and scanned books, as well as excerpts from commercial books, provided as previews by publishers. The sister webpage is *Google Scholar* (►http://scholar.google.com), which was introduced in 2004 and searches online journals rather than books. The idea behind this project was to provide a freely-available article search engine as an alternative to some of the commercial products such as Elsevier's *Scopus* or Clarivate Analytics' *Web of Science*.

The Web of Science was founded by Eugene Garfield in 1960, subsequently acquired by the Thomson Reuters, an information company headquartered in New York City (USA) and Toronto (Canada), and in 2016 by Clarivate Analytics, a independent company running several subscription-based services in various fields. The Web of Science can be accessed by universities and research institutions from the *Web of Knowledge* (►http://apps.webofknowledge.com) index. This index includes more than one billion searchable, cited references. It also includes the InCities Journal Citation Reports, offering a systematic, objective means to evaluate the world's leading journals, according to the site's webpage.

A commercial bibliographic data base specializing in earth science literature is *GeoRef* (►http://www.agiweb.org/georef) produced by the American Geological Institute (AGI). GeoRef was launched in 1966 as a series of printed catalogues, later evolving into a CD-based system and ending up as an online service. According to the AGI webpage, as of October 2016, data base contains over 3.7 million references including journal articles, books, maps, conference papers, reports, and theses.

2.3 Finding the Relevant Literature

We now use the Internet resources introduced in the previous section to demonstrate a typical search for books and journal articles, using various indices to rank the literature. We first start with commercial online bookstores, proceed to Google Books and Google Scholar, and end up at professional literature search data bases such as the Web of Science.

2.3.1 Searching with Online Bookstores

In our first example, we would like to explore a new field for future research activities, e.g., human evolution. Let us assume that we are planning to spend €50 on relevant books and start with a literature search on *Amazon* (►http://amazon.com) or any other online bookstore. After entering *human evolution* in the search field, we obtained a list of 24,641 books, DVDs and other products, showing the results in order of relevance. A search for *human evolution* in October 2016 produced the following list of books (◘Fig. 2.1):

```
1. Harari YN (2015) Sapiens: A Brief History of Humankind. Harper

2. Stringer C (2013) Lone Survivors: How We Came to Be the Only Humans on
Earth. St. Martin's Griffin

3. Stringer C, Andrews P (2005) The Complete World of Human Evolution.
Thames & Hudson
```

2

```
4. Roberts A (2011) Evolution: The Human Story. DK

5. Wood B (2005) Human Evolution: A Very Short Introduction (Very Short
Introductions). OUP Oxford

6. Harris EE (2014) Ancestors in Our Genome: The New Science of Human
Evolution. Oxford University Press

7. Pyne L (2016) Seven Skeletons: The Evolution of the World's Most Famous
Human Fossils. Viking

8. Lieberman D (2014) The Story of the Human Body: Evolution, Health, and
Disease. Vintage

9. Tattersall I (2015) The Strange Case of the Rickety Cossack: and Other
Cautionary Tales from Human Evolution. St. Martin's Press

10. Hands J (2016) Cosmosapiens: Human Evolution from the Origin of the
Universe. The Overlook Press
```

The search technology is provided by *A9* (►http://a9.com), which is owned by Amazon. Amazon also ranks all products by their sales from their own webpages (the *Amazon sales rank*, or ASR, which is updated every hour). Sales, not citations, make a book popular at Amazon: both professionals working in the field of human evolution and interested lay people contribute to the sales of a book, and therefore to the ranking of

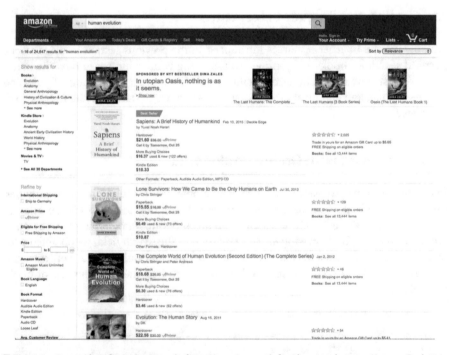

☐ **Fig. 2.1** Screenshot from Amazon (►http://amazon.com) for the search term *Human Evolution*. The date of screenshot is 24 October 2016, whereas the search experiment described in the text was performed in early 2011, which explains the slight difference in search results

a particular product. What makes a customer decide to buy a particular book through Amazon? Why are certain books popular at Amazon and others not? There is much speculation about the exact mathematical algorithm used by the company to calculate the ranking of a book. The obvious factors that may increase the sales of a book are a low price, recent publication date, good title, attractive cover, great illustrations, and positive customer reviews. The reputation of the author is certainly an important criterion within the professional community but might be a secondary consideration for other customers.

Yuval Noah Harari, born in 1976, author of the first-ranked book *Sapiens: A Brief History of Humankind* and author of four other books, is a professor of history at the Department of History, Hebrew University, Jerusalem. He has a relative short list of twelve scientific articles, with a total number of citations of only 42 (October 2016) and a relatively low *h*-factor of 4, according to the *Web of Science* results, suggesting that popular writing interests him more than scientific work (►http://www.ynharari.com). In contrast, Bernard Wood, who authored the fifth-ranked book entitled *Human Evolution: A Very Short Introduction* is an anthropologist and the University Professor of Human Origins at the George Washington University. His paperback book of 148 pages, which provides a very brief summary of current knowledge concerning human origins, received mixed but predominantly positive reviews. Chris Stringer from the Natural History Museum in London, who authored the second-ranked and, together with Peter Andrews from the same institution, third-ranked book, is also an active researcher in the field of human evolution and has, similar to Peter Andrews, authored numerous highly-cited journal articles and books. Both books are popular science books by active researchers, as the short description of the books suggests. In contrast, Chris Stringer's other book with the same coauthor, *Principles of Human Evolution*, published in 2003, states very clearly, that it is a textbook for students and professionals rather than a popular science book for an interested lay audience. The book was tenth-ranked at Amazon in early 2011 but no longer appears in the list, probably because there are newer books on the topic. Robert Lewin also authored the book *Human Evolution: An illustrated Introduction*, which is another popular book written by an active researcher in the field of paleoanthropology. A very good example of a popular book written by an expert is the book

Potts R, Sloan C (2010) What Does It Mean to be Human? National Geographic

Rick Potts is the director of the Smithsonian Institution's Human Origins Program (►http://humanorigins.si.edu), and Chris Sloan is the National Geographic's expert in paleoanthropology (►http://www.nationalgeographic.com). In this example, the collaboration between an active researcher and a popular science writer has resulted in an excellent book for both experts and lay persons. These examples illustrate that authors of science books can range from science writers and journalists with a science background, often within the field of the book's topic, to active researchers writing both popular science books for an educated lay audience and textbooks for students or professionals with a particular interest in the topic.

The books *The Complete World of Human Evolution (Second Edition) (The Complete Series)* by Stringer and Andrews, *Human Evolution: A Very Short Introduction* by

2

Wood and *What Does It Mean to be Human?* by Rick Potts together come to less than €50 and cover the full range of book types, from a textbook for students to a popular book for the interested lay person. As demonstrated in this section, however, it may require a lot of background information on the various authors before the best books on a specific topic can be selected. The next subsections provide some more objective criteria with which to rank books and journal articles.

2.3.2 Searching with Google Books and Google Scholar

What alternative rankings are available that are not based on commercial factors and undisclosed mathematical algorithms? Let us try *Google Books*, as a non-commercially biased search service. By once again, entering *human evolution* in the search field, we obtained a list of about 2.2 million results in early 2011 sorted by relevance. Here are the top ten items on the Google Books list:

```
1. Tuttle RH (2014) Apes and Human Evolution. Harvard University Press

2. Taylor T (2010) The Artificial Ape: How Technology Changed the Course of
Human Evolution. Palgrave Macmillan

3. Muckle RJ, Tubelle de González L (2015) Through the Lens of Anthropology:
An Introduction to Human Evolution and Culture. University of Toronto Press

4. Bradshaw JL (2014) Human Evolution, A Neuropsychological Perspective.
Psychology Press

5. Wood B (2005) Human evolution: a very short introduction. Oxford
University Press

6. Lewin R (2005) Human Evolution: An Illustrated Introduction.
Wiley-Blackwell

7. Ehrlich PR, Ehrlich AH (2012) The Dominant Animal: Human Evolution and
the Environment. Islandpress / Shearwater Books

8. Regal B (2004) Human Evolution: A Guide to the Debates. ABC-CLIO

9. Barnes E (2007) Diseases and Human Evolution. University of New Mexico
Press

10. Campbell BG (2009) Human Evolution: An Introduction to Mans Adaptations,
4th Edition. Aldine Transaction
```

This certainly is a significantly different type of list, which may not only include books but also book chapters; it is computer generated by an automated search for keywords without the involvement of any quality criteria. Some of the books discussed above also appear on the Google Books list, the other items listed are relatively less popular books or monographs on specific aspects, current debates, or perspectives of human evolution. This example demonstrates that, although biased by non-scientific criteria, searching for books on the basis of sales-based rankings seems to be the better option for finding the relevant literature.

Finding relevant journal articles on a specific topic is a much simpler task by far. In this case, the rankings are based on citations by experts rather than on sales to a mixed professional and non-professional clientele. As an example, we aim to find scientific literature on the influence of the Younger Dryas cold reversal in Africa. The online encyclopedia provides some general information on the Younger Dryas as a climate event:

```
http://en.wikipedia.org/wiki/Younger_Dryas
```

Authored by anonymous and possibly non-expert contributors, this article states:

```
The Younger Dryas is a geological period from c. 12,900 to c. 11,700 calendar
years ago (BP). It is named after an indicator genus, the alpine-tundra
wildflower Dryas octopetala. Leaves of Dryas octopetala are occasionally
abundant in the Late Glacial, often minerogenic-rich, like the lake sediments
of Scandinavian lakes. The Younger Dryas saw a sharp decline in temperature
over most of the northern hemisphere, at the end of the Pleistocene epoch,
immediately preceding the current warmer Holocene.
```

The article continues for about three hundred words with information on the causes and consequences of the Younger Dryas period, before the first reference occurs,

```
Carlson, A. E. (2013). "The Younger Dryas Climate Event" (PDF). Encyclopedia
of Quaternary Science. 3. Elsevier. pp. 126-34.
```

which is a comprehensive six-page encyclopedia article on the Younger Dryas. The small number of references in a introductory chapter is certainly not usual in scientific essays. Also in the following sections are references rare, for which Wikipedia was often criticized. The section about the timing of the onset and termination of the Younger Dryas cites

```
Alley, Richard B. (2000). "The Younger Dryas cold interval as viewed from
central Greenland". Quaternary Science Reviews. 19 (1): 213-226. doi:10.1016/
S0277-3791(99)00062-1.
```

We are not sure whether this article is one of the most important articles on this subject. The articles on Wikipedia are, however, always a good starting point for an in-depth literature research on a specific topic. Knowing that the Younger Dryas, in general, is a time interval between 12,900 and 11,700 years ago that was characterized by a cold climate at least points us in the right direction for our literature review on the topic.

We next use *Google Scholar* (▶http://scholar.google.com) for a more detailed search of journal articles (◻Fig. 2.2). For this we enter *Younger Dryas Africa* in the search field and get a list of about 12,800 references. According to Google Scholar's webpage, it aims to rank documents the way researchers do, weighing the full text of each document, where it was published, who it was written by, as well as how often and how recently it has been cited in other scholarly literature. Browsing the list of articles retrieved by our search, however, it is not obvious how the articles are ranked, but the

2

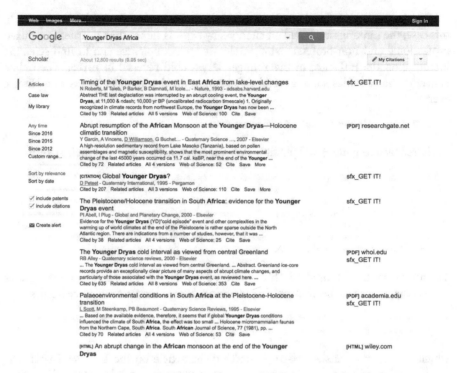

■ **Fig. 2.2** Screenshot from the *Advanced Scholar Search* of *Google Scholar* (▶ http://scholar.google.com), taken on 24 October 2016. Google Scholar was introduced in 2004 and included online journals rather than only books

number of citations and the occurrence of the search terms in the title certainly seem to have had some influence. The top ten articles on the list are:

1. N Roberts, M Taieb, P Barker, B Damnati, M Icole, D Williamson (1993) Timing of the Younger Dryas event in East Africa from lake-level changes. Nature 366:146-148 [Citations: 139]

2. Y Garcin, A Vincens, D Williamson, G Buchet, J Guiot (2007) Abrupt resumption of the African Monsoon at the Younger Dryas-Holocene climatic transition. Quaternary Science Reviews 26:690-704 [Citations: 72]

3. D Peteet (1995) [CITATION] Global Younger Dryas? Quaternary International 28: 93 [Citations: 207]

4. PI Abell, I Plug (2000) The Pleistocene/Holocene transition in South Africa: evidence for the Younger Dryas event. Global and Planetary Change 26:173-179 [Citations: 38]

5. RB Alley (2000) The Younger Dryas cold interval as viewed from central Greenland, Quaternary Science Reviews 19:213-226 [Citations: 635]

6. L Scott, M Steenkamp, PB Beaumont (1995) Palaeoenvironmental conditions in South Africa at the Pleistocene-Holocene transition. Quaternary Science Reviews 14:937-947 [Citations: 70]

```
7. MR Talbot, ML Filippi, NB Jensen, JJ Tiercelin (2007) An abrupt change in
the African monsoon at the end of the Younger Dryas. Geochemistry Geophysics
Geosystems 8:Q03005 [Citations: 38]

8. D Rind, D Peteet, W Broecker, A McIntyre, W Ruddiman (1986) The impact of
cold North Atlantic sea surface temperatures on climate: implications for the
Younger Dryas cooling (11-10 k). Climate Dynamics 1:3-33 [Citations: 251]

9. J Ortiz, T Guilderson, J Adkins, M Sarnthein ... (2000) Abrupt onset and
termination of the African Humid Period: rapid climate responses to gradual
insolation forcing. Quaternary Science Reviews 19:347-361 [Citations: 921]

10. P Fawcett, AM Ágústsdóttir, RB Alley ... (1997) The Younger Dryas
termination and North Atlantic Deep Water formation: Insights from climate
model simulations and Greenland ice cores. Paleoceanography 12:23-38
[Citations: 105]
```

Depending on our institution's journal subscriptions, we may be able to directly access the PDFs of these papers from the publisher's website by following the link provided in the Google Scholar list of articles. As we will see later, the first article, which is by N. Roberts, was indeed the earliest publication on the influence of the Younger Dryas cold event in Africa. Published in *Nature* in 1993, it was the result of a French-British collaboration in the 1980s and 1990s led by M. Taieb, and is fairly-well cited with 74 citations. The paper by Y. Garcin is one of many follow-up publications from this project, although the main conclusions are different from those in the Roberts paper (see ▸Sect. 2.4). Articles no. 5, 8, and 10 on the list provide a more global perspective on the influences of the Younger Dryas, including some comments on Africa. The highly cited article no. 9 provides some insights into the onset and termination of the Younger Dryas in Africa, but the first author Peter deMenocal is actually missing in the reference. The correct references is

```
9. P deMenocal, J Ortiz, T Guilderson, J Adkins, M Sarnthein, L Baker,
M Yarusinsky (2000) Abrupt onset and termination of the African Humid Period:
rapid climate responses to gradual insolation forcing. Quaternary Science
Reviews 19:347-361 [Citations: 921]
```

As we can see from this experiment, our Google literature search gives some reasonable results but with limitations.

2.3.3 Searching with the Clarivate Analytics Web of Science

There are not many possibilities for influencing the search results in Google Scholar. The exact routine that ranks the journal articles for their relevance is not obvious. In our next experiment, we use the commercial Clarivate Analytics *Web of Science* (▸https://apps.webofknowledge.com) literature search data base, which is very similar to Elsevier's Scopus service (▸https://www.scopus.com). For this we enter "*Younger Dryas*" "*East Africa*" (with double quotes) in the *topic* search field. In October 2016 this yielded a list of 44 articles, ranked by publication date (◘Fig. 2.3). In contrast to Google Scholar, the Web of Science allows changes to be made to the mode of ranking, which can be, e.g., by the first author, the publication date, the number of citations, and the title of the journal. There is also the option to sort by relevance using a

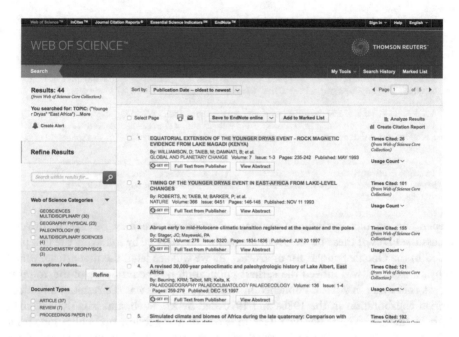

■ **Fig. 2.3** Screenshot from Clarivate Analytics Web of Knowledge (▶http://isiknowledge.com), showing results of a search for *Younger Dryas Africa*. The date of the screenshot is 24 October 2016

ranking system that considers how many of the search terms are found in each record, according to *Help*. Sorting by relevance, the article

```
Effect of aridity and rainfall seasonality on vegetation in the southern
tropics of East Africa during the Pleistocene/Holocene transition
By: Ivory, Sarah J.; Lezine, Anne-Marie; Vincens, Annie; et al.
QUATERNARY RESEARCH Volume: 77 Issue: 1 Pages: 77-86 Published: JAN 2012
```

which has only been cited 14 times during the last decade, ranked first in October 2016. Sorting the list of 44 articles on the Younger Dryas in East Africa by the number of citations, the article

```
Simulated climate and biomes of Africa during the late quaternary:
Comparison with pollen and lake status data
By: Jolly, D; Harrison, SP; Damnati, B; et al.
QUATERNARY SCIENCE REVIEWS Volume: 17 Issue: 6-7 Pages: 629-657 Published: 1998
```

ranks first, with 192 citations. Sorting by publication date, the article

```
EQUATORIAL EXTENSION OF THE YOUNGER DRYAS EVENT - ROCK MAGNETIC EVIDENCE
FROM LAKE MAGADI (KENYA)
By: WILLIAMSON, D; TAIEB, M; DAMNATI, B; et al.
GLOBAL AND PLANETARY CHANGE Volume: 7 Issue: 1-3 Pages: 235-242 Published:
MAY 1993
```

which was ranked first in Google Scholar, is the oldest article of relevance on the Younger Dryas in Africa. Since the article has more than three authors, the co-authors were merged as *et al.*, which stands *et alii*, Latin for *and others*. The exact format, in particular the use of et al., depends on the format required for citations by the particular journal. Clicking on the number *26* opens a new webpage listing the 26 articles citing *Williamson et al.* (1993). After having identified this article as one of the earliest to provide evidence of a Younger Dryas influence in Africa, we also find

```
TIMING OF THE YOUNGER DRYAS EVENT IN EAST-AFRICA FROM LAKE-LEVEL CHANGES
By: ROBERTS, N; TAIEB, M; BARKER, P; et al.
NATURE Volume: 366 Issue: 6451 Pages: 146-148 Published: NOV 11 1993
```

which was published in the same year. What other criteria would help us to identify the best article on the topic, rather than relying on simple counts of the number of times the search term "*Younger Dryas*" "*East Africa*" is found in a record? In addition to the quality of the content, three measures are available to assist us in finding the best article on the topic; these measures are widely used but their validity is intensely debated.

The first measure for the relevance of an article is the *number of citations*, already mentioned when we used Google Scholar to search for literature. The number of citations counts all citations of an article in journals listed on the Web of Science since the article's publication date. It therefore does not count the citations of an article in books or other media. In this regard the article by N. Roberts is more frequently cited (101 times based on Web of Science, 140 times according to Google Scholar) than the article by D. Williamson (26 times). On the basis of this measure, the article by N. Roberts therefore appears to be more relevant, but let us explore some other criteria before we come to a final decision on the most relevant publication.

The second measure for the relevance of an article is the journal *impact factor* (IF). The journal impact factor can be obtained from the *Journal Citation Reports* (JCR) of the Web of Science. The JCR lists the number of articles published in the journal per year, the number of citations of that journal in articles published during the same year, the impact factor calculated from that year, and many other parameters. The IF calculated annually is the average number of citations of an article in journals during the two previous years. The IF measures only citations for journals that are listed on the Web of Science and also only counts citations from those journals.

Since being introduced by Eugene Garfield, the founder of the Science Citation Index (SCI), the impact factor has been under severe criticism. For instance, the IF is highly dependent on the discipline. The journal *Nature*, in which the article by N. Roberts was published, and its subsidiary journal *Nature Geoscience*, are good examples of this dependency. The British journal *Nature* and its American counterpart *Science*, are the world's most respected scientific journals, with 2015 impact factors of 38.138 (Nature) and 34.661 (Science), i.e., articles in these journals were, on average, cited about 35–38 times during 2013 and 2014. The impact factor of these and other journals can be accessed from the Web of Science webpage by clicking the *Journal Citation Reports* (JCR) tab. There you can select the JCR edition and year, e.g., the *JCR Science Edition* (SCIE) for the year *2015*, and enter *Nature* in the *Go to Journal Profile* field and you will get a table with the rank of the journal based on the IF, the total number of citations, the IF, and the five-year IF, as well as many other parameters for measuring the success of a journal.

Going back to the article by N. Roberts, cited 101 times since its publication date, we can analyze the citation record over the last two decades. Again, clicking on the number *101* (the number of citations) opens a new webpage listing all other articles citing *Roberts et al.* (1993). There we click on *Analyze Results* and get a new webpage with various options for ranking and sorting the results. We then choose *Publication Year* to rank the records and *Selected field* to sort the results. This yields a chart displaying the citations per year where we can see that the article typically receives 5 to 7 citations per year, with a peak of 14 citations in the year 2000. Note that this citation rate is significantly lower than the journal's two-year average citation rate, i.e., its impact factor. However, Nature's citation rate in 1993 was different from that of today. The article by N. Roberts, which certainly made an important contribution to the understanding of the influence of high-latitude climate change in the tropics and provided the first evidence of the Younger Dryas cold event in East Africa, does not receive a large number of citations simply because it covers a relatively restricted field of research in which a low number of active scientists are involved.

The article's maximum citation rate, however, is within the range of the IF of Nature Geoscience (IF 12.508), which was launched as a disciplinary sub-journal of Nature in January 2008. H. Langenberg, editor at Nature Geoscience, explained the difference between the IFs of the Nature and Nature Geoscience as being due to the average citation rates per paper varying by a factor of five or six, depending on the scientific field, with the geosciences at the low end compared to other fields such as genetics and stem-cell research in which articles receive many more citations (Langenberg 2010). The *Proceedings of the National Academy of Sciences of the United States of America* (PNAS) is another interdisciplinary journal publishing original research; it has a 2015 IF of 9.423, and is therefore ranked slightly lower than Nature Geoscience, which is the most highly ranked specialized journal in our field.

Other highly ranked journals in earth sciences are *Reviews of Geophysics* (IF 11.444) and *Earth-Science Reviews* (IF 6.991). Below these, numerous journals are ranked with an IF in the 3–5 range, such as *Geology* (IF 4.548), *Quaternary Science Reviews* (IF 4.521), *Earth and Planetary Science Letters* (IF 4.326) and the more specialized journal *Paleoceanography* (IF 3.433). Almost 90% of all earth science journals, however, fall within the IF>3.000 range, presumably including articles that are not cited at all. The article by D. Williamson was published in *Global and Planetary Change*, a journal ranking at the lower end of the top 10% of journals according to its IF of 3.548. Based on the IF of the two journals in which the articles about the Younger Dryas in Africa were published, the article by N. Roberts is again the better choice.

The third measure, which describes publication success, is the *h-factor* of the first author. This parameter, which is even more intensively debated than the previous two, was introduced by J.E. Hirsch to quantify an individual's scientific research output (Hirsch 2005). The index h is defined as the number of papers published by a scientist with citation number $\geq h$. According to J.E. Hirsch, the index h for a given individual should increase approximately linearly with time, i.e., scientists produce papers of similar quality at a steady rate over the course of their careers. Of course, the slope m of h over n years varies between different researchers, again depending not only on publication success but also on the discipline and on co-authorship regulations. On the basis of the slope m, J.E. Hirsch concludes that $m \approx 1$, i.e., $h = 20$ after 20 years, characterizes a successful scientist, $m \approx 2$ characterizes an outstanding scientist, and $m \approx 3$ characterizes truly unique individuals.

To determine the *h*-factor for the two first authors, N. Roberts and D. Williamson, we click on the *Web of Science* tab on the Web of Knowledge webpage, and then enter *Roberts N* in the *Author* search field. From the *More Settings* menu, we enable the *Science Citation Index Expanded* data base and disable the social sciences and arts and humanities citation indices, and then click *Search*. This yields a list 1,942 references (October 2016), including those for other scientists with the same name as the author of the article on the Younger Dryas. We can *Select articles grouped for author name* by clicking on *Roberts N* on the left to exclude those authors from other fields such as clinical sciences, biology of reproduction and librarianship, and keep the person authoring articles in the field of physical sciences. Listed as no. 6, we find *ROBERTS N*, also published as *ROBERTS NEIL* and *ROBERTS NJ,* from the University of Plymouth, and science articles in the field of Geology (36), Physical Geography (22), Ecology (15), Marine Freshwater Biology (8) and Paleontology (6) in the years of 1979–2015.

This selection reduces the total number of publications to 58, ranking the Nature article five with 101 citations. We can a citation report by clicking *Create Citation Report* on the left. According to this report, the total number of citations of N. Roberts is 1917, the average number of citations per item is 36.87 and the *h*-factor is 25. Since we are not sure whether we have indeed included all publications by N. Roberts, we return to the *Results* page and *Refine Results* by excluding *Web of Science Categories.* We get a long list of subject areas, in which we activate the field of physical sciences, which includes geosciences, geography, geochemistry and geophysics, environmental sciences, multidisciplinary sciences, limnology, paleontology, oceanography, and archeology, and then click *Refine.* This again reduces the number of publications, but it soon becomes evident that we have again included authors with similar names but from different fields, such as N. Roberts working on environmental trace-metal contamination.

Excluding those publications from authors such as Noel Roberts, Nicholas J. Roberts and others, we end up with pretty much the same number of about 58 publications by the C. Neil Roberts that works at the University of Plymouth and led projects on paleoenvironmental change in Turkey after he stopped working in Africa, as we had before. Cross-checking the Journal Citation Reports with the (incomplete) list of publications provided on his webpage (►http://www.plymouth.ac.uk/staff/cnroberts), we are very confident that we have arrived close to the true publication record of the N. Roberts authored the article on the Younger Dryas in Africa. As I know from the author, however, both our search results and Neil Roberts' webpage are missing one important article, which was first-authored by Henry Lamb from the University of Aberystwyth and co-authored by N. Roberts:

    Lamb, H.F., Gasse, F., Benkaddour, A., el-Hamouti, N., van der Kaars,
    S., Perkins, W.T., Pearce, N.J., Roberts, C.N. (1995) Relation between
    century-scale Holocene arid intervals in tropical and temperate zones,
    Nature, 373, 134-137.

According to the Web of Science, this article has been cited 171 times. This article is missing from our search list because in the list of authors Roberts is listed as *C.N. Roberts.* Neil Roberts is known by his middle name Neil, but some publications, such as the one published in Nature, also add his first initial C. Note that the same author has also published the very popular textbook

```
Roberts N (2014) The Holocene: an environmental history - 3rd edition.
Wiley-Blackwell
```

which is not mentioned at all in the Citation Reports. Not considering books is of course a significant weakness of the Web of Science. Books do not contribute to any of the measures of the scientific or publication success of a researcher. Recently, however, Thomson Reuters has launched a *Book Citation Index* for the Web of Knowledge, now owned by Clarivate Analytics, which can help to find the most relevant books on a topic, using similar parameters to those used for journal articles to measure the contribution that books make a particular discipline.

How does D. Williamson's h-factor compare with that of N. Roberts? The first search returns about 2,701 references, but can *View Distinct Author Sets* reduce the articles to only those published by the French researcher D. Williamson? By including only publications related to geosciences we reduce the list to about 63 articles, with a total of 1,600 citations, an average citation rate of 25.40, and an h-factor of 23. D. Williamson, who led the paleomagnetics laboratory at the CEREGE in Aix-en-Provence, benefitted from being a laboratory manager. He is co-author of many highly-cited paleoclimate publications that used paleomagnetics, but also first-authored several well-cited articles on mineral-magnetic proxies in paleoclimate research. D. Williamson is currently at the Institut de Recherche pour le Développement (IRD) in Marseille, France.

Please note that he is also a co-author on N. Roberts' Nature paper, to which he probably contributed the magnetics record shown in Figure 2 of the article. Comparing the authors listed for both papers

```
Roberts N, Taieb M, Barker P, Damnati B, Icole M, Williamson D (1993) Timing
of the Younger Dryas event in East Africa from lake-level changes. Nature
366:146-148

Williamson D, Taieb M, Damnati B, Icole M, Thouveny M (1993) Equatorial
extension of the Younger Dryas Event - Rock Magnetic Evidence From Lake
Magadi (Kenya). Global and Planetary Change 7:235-242
```

reveals that other authors are also listed for both articles. The Nature article, however, is missing in our search results for D. Williamson, and we are therefore sure that our publication record is not complete. An Internet search for some of these authors leads to the conclusion that both N. Roberts and D. Williamson worked under the guidance of Maurice Taieb, who was the leader of the paleolimnological project at Lake Magadi in Southern Kenya. M. Taieb is a French geologist and paleoanthropologist who worked in Africa for over fifty years, and was involved in the discovery of the fossil fragments of *Lucy* in the early 1970s.

Because of the leadership and co-authorship of M. Taieb on both publications, and because D. Williamson's article is on the paleomagnetic record only whereas N. Roberts' paper includes a discussion of other climate proxies such as diatoms (silica algae), organic carbon, and calcite, and also because Nature is a far better reference than Global and Planetary Change, we come to the conclusion that the article by

N. Roberts is, in general, likely to be the more useful publication on the Younger Dryas in Africa. An analysis of our literature research, however, clearly shows how difficult it is to evaluate the relevance of a specific article on the basis of bibliometric measures such as journal impact factors, citation indices, and *h*-factors. As the demonstration clearly shows, it often requires a lot of background knowledge to find the most relevant papers on a particular topic. Of course having found the most important journal article should not prevent you from also reading some of the less relevant papers. These may well contain interesting details in the text that are more relevant to your own particular research than to that of the citing authors.

2.4 Extracting the Relevant Information from Literature

Having identified N. Roberts' article as the most relevant reference to use as a starting point for research into influence of the Younger Dryas event in East Africa, we now obtain the electronic version of this article, which was published in Nature (►http:// www.nature.com) in 1993. The journal's webpage has a search field in which we can enter the terms *Younger Dryas East Africa* and get about 40 references as a result. We see that there are even older articles mentioning the possible influence of the Younger Dryas in Africa based, for example, on analysis of the sapropels in the Mediterranean Sea as indicators of monsoonal rainfall intensity:

```
Rossignol-Strick M, Nesteroff W, Olive P, Vergnaud-Grazzini C (1982)
After the deluge: Mediterranean stagnation and sapropel formation. Nature
295:105-110
```

The article by N. Roberts also shows up in the 40 references, listed as a *Letter to Nature* type of article. A Letter in Nature is four pages long and has no more than 30 references, whereas an *Article* can be up to five pages long and has up to 50 references; *Brief Communications* are even shorter than Letters.

```
Timing of the Younger Dryas event in East Africa from lake-level changes
Nell Roberts, Maurice Taieb, Philip Barker, Brahlm Damnati, Michel Icole,
David Williamson
Nature 366, 146-148 (11 November 1993) doi:10.1038/366146a0 Letter
Abstract | PDF | Rights and permissions | Save this link
```

Please note that the author's first name Neil is actually misspelled Nell in the journal's article data base (November 2016). The person who typed the reference into the data base or used a optical character recognition, presumably used the scanned version of the article, which we download later from the journal's webpage, in which the first name indeed looks like Nell rather than Neil.

As an alternative, we can browse the contents of the journal if we know the volume and page numbers for the article. On the main page of Nature, we click on *Journals A-Z* under the *MENU* button on the left. There we choose the journal *Nature* from the list of publications, then *Archive* from the *Menu* of the journal's webpage, select the year *1993*, volume *11 November 1993 366 6451 95-188* and get the contents list for the issue of the journal that contains the article by N. Roberts.

2

Timing of the Younger Dryas event in East Africa from lake-level changes 146
NELL ROBERTS, MAURICE TAIEB, PHILIP BARKER, BRAHLM DAMNATI,
MICHEL ICOLE & DAVID WILLIAMSON
doi:10.1038/366146a0
First paragraph & References | PDF (355K)

The reference contains the title, the page number (*146*), the authors, the digital object identifier (*doi:0.1038/366146a0*), the first paragraph and references, and a link to the full article. The *digital object identifier* (DOI) is a unique and persistent identifier of electronic documents, such as books, journal articles and data sets. The title, list of authors, abstract and references, are freely available without subscription to the journal (◻Fig. 2.4). The full article, however, is available by subscription only. In many cases, however, the title and the abstract already contain the most important information from the article. We are interested to learn about the influence of the Younger Dryas in East Africa. If the Younger Dryas did have an influence, how did it affect the climate in East Africa and when? The title

Timing of the Younger Dryas event in East Africa from lake-level changes

suggests that there was indeed an influence of the Younger Dryas in East Africa, and that this influence has been identified in the water-level record of Lake Magadi in Kenya. Furthermore, the title indicates that the article discusses the timing of the

◻ **Fig. 2.4** Screenshot from *Nature* (▶http://nature.com) showing the freely-available title, list of authors, abstract, and references from the paper by Roberts et al. (1993). The date of the screenshot is 24 October 2016

Younger Dryas influence in the region. It does not, however, contain information on the nature of the influence (e.g., whether it was resulted in a drier or wetter climate) or the exact timing of the resulting climate change in East Africa.

We proceed to the abstract of the article, which typically provides a summary, in most cases without reference to previous articles by other researchers. For articles in Nature, however, the abstract is a fully-referenced first paragraph, in bold format.

```
THE last deglaciation was interrupted by an abrupt cooling event, the
Younger Dryas, at 11,000-10,000 yr BP (uncalibrated radiocarbon timescale)1.
Originally recognized in climate records from northwest Europe, the Younger
Dryas has now been identified in marine and ice-core records worldwide2-6.
In the tropics, a broadly contemporaneous change in climate is recorded by
decreases in water levels and increased salinity of lakes7-9,14, indicating
a period of arid climate caused by a reduction in ocean-to-land moisture
flux. The exact timing of these changes in relation to the Younger Dryas
event in high-latitude records has remained unclear, however. Here we
present climate records based on analyses of diatom assemblages, geochemistry
and magnetic mineralogy of radiocarbon-dated sequences of laminated lake
sediments from Lake Magadi in the East African rift. These records provide
a detailed record of climate change in lowland equatorial Africa throughout
the last deglaciation (12,800-10,000 14C yr BP). We find that lake-level
and humidity maxima coincide with the most rapid phases of ice melting in
the Northern Hemisphere, and that the climate changes, including the Younger
Dryas event, were synchronous at low and high latitudes. Thus, the effects
of abrupt climate change appear to be felt at both high and low latitudes
without a significant time lag.
```

The small superscript letters refer to articles in the list of references. As an example, reference 1 after the first sentence refers to

```
1. Zahn, R. Nature 356, 744-746 (1992).
```

which is the article by R. Zahn

```
Zahn, R (1992) Deep ocean circulation puzzle. Nature 356:744-746
```

on the Younger Dryas event in general. This article is cited after the first sentence of the article by N. Roberts. This sentence introduces the phenomenon of the Younger Dryas as an abrupt cooling event between 11,000 and 10,000 ^{14}C yr BP during the last deglaciation, i.e., during the transition from the last glacial to the present interglacial. The second and third sentences state that the Younger Dryas was a global event that resulted in a drier climate in the tropics. The exact timing of tropical climate change during the Younger Dryas event, however, remains unclear.

Following an outline of the scientific question to be investigated, the second half of an abstract or first paragraph typically starts with the statement *Here we present ...*, introducing the approach, results, and conclusions of the actual research presented in the article. In our example, this reads as follows:

```
Here we present climate records based on analyses of diatom assemblages,
geochemistry and magnetic mineralogy of radiocarbon-dated sequences of
laminated lake sediments from Lake Magadi in the East African rift.
```

from which the reader learns that the article is about a lake-sediment record from a lake located in the East African rift, and that it includes the analysis of fossil silica

2

algae (diatom) assemblages, as well as chemical and physical sediment parameters. The last three sentences of the abstract

> These records provide a detailed record of climate change in lowland equatorial Africa throughout the last deglaciation (12,800–10,000 14C yr BP). We find that lake-level and humidity maxima coincide with the most rapid phases of ice melting in the Northern Hemisphere, and that the climate changes, including the Younger Dryas event, were synchronous at low and high latitudes.

basically list the results—more specifically, the age of the sediments and therefore the time interval contained in the lake-level record, and the information that the climate change coincided with the high-latitude deglaciation—but it does not contain precise information on precisely how the Younger Dryas affected the climate in the region. Remembering that the first part of the paragraph said that the Younger Dryas was generally dry in the tropics and reading here that humidity maxima coincided with the ice melting, we could conclude that the Younger Dryas cold event was probably dry. The ultimate conclusion of the article is then given in the final sentence of the abstract

> Thus, the effects of abrupt climate change appear to be felt at both high and low latitudes without a significant time lag.

essentially proposing a link between high and low latitudes based on the correlation of climate records without any time lag.

Reading the title and the abstract (or the first paragraph) of the article has taken us only one or two minutes, but we have already extracted the most important scientific information, despite the minor weakness in this particular article of not being precise about the exact nature of the influence that the Younger Dryas had in East Africa. In our experiment this weakness necessitates an analysis of the full article, or at least parts of it. The article is in a *Portable Document Format* (PDF). This format was designed by Adobe Systems (▶http://www.adobe.com) as self-contained cross-platform document. PDF files contain the complete formatting of vector illustrations, raster images and text, or a combination of all these, and includes all necessary fonts. These files are highly compressed, allowing a fast Internet download. Adobe Systems provides the *Acrobat Reader* free-of-charge for all computer platforms, for reading PDF files (▶https://get.adobe.com/reader). As we will see later, more recent articles are in a vector format, allowing text fragments and graphics to be extracted and incorporated into lectures, or into books such as this, provided that the copyright of the electronic materials is respected. The article by N. Roberts, published in 1993, is a scanned version of the original printed journal article and therefore only available in raster format.

The article is approximately 2.5 pages long and therefore represents are relatively short piece of text. Nature, Science and the Proceedings of the National Academy of Sciences (PNAS) typically publish articles with a maximum length of four pages, whereas most other journals allow papers of 10 printed pages, or even more. Most journals also allow the authors to publish an electronic supplement to the printed article, including more text, tables, figures, references and multimedia. Reading an article of such length requires efficient reading, and we therefore need to know the most likely location of the most relevant information in an article. Fortunately, research articles share a common structure comprising the title and abstract, introduction, setting and methods, results, discussion, and conclusions. Differences exist between journals, however, and Nature, Science, and PNAS each structure articles in a slightly different way. The overall structure of the introduction, methods, results, and discussion/conclusions also applies to these journals but the sections do not have headers. Details about the geographic and geologic setting of the study area, an overview of the methods used in the analysis, and results presented as tables and text are usually published separately from the actual article as electronic supplementary information.

Apart from the information contained in the title and abstract/first paragraph, the most important information typically occurs in the conclusion or the last paragraph of the article, and in the last figure. The last figure of an article often summarizes the results or places a new data set in a larger context by, for instance, correlating the new data with established records of past climate change, or with any other data set from a different location and from a different research project. In our example, Figure 3 compares the water-level record from Lake Magadi in Kenya with other climate records, namely the oxygen-isotope curve from Rotsee in Switzerland by Lotter et al. (1989) and winter temperatures for Britain reconstructed from beetle remains by Atkinson et al. (1987):

```
Lotter A, Zbinden H (1989) Late-Glacial pollen analysis, oxygen-isotope
record, and radiocarbon stratigraphy from Rotsee (Lucerne), Central Swiss
Plateau. Ecologae geol. Helv. 82:191-202

Atkinson TC, Briffa K, Coope GR (1987) Seasonal temperatures in Britain
during the past 22,000 years, reconstructed using beetle remains. Nature
325:587-592.
```

The figure clearly shows that the water level dropped by about 40 m during the Younger Dryas, suggesting that tropical East African climate became drier when the oxygen isotopes and beetle assemblages indicate colder temperatures in the mid latitudes. The relevant text in the last paragraph, which typically contains the conclusion of the article, also says that

```
Conditions became warmer and wetter around 12,700 14C yr BP, cooler and
drier at 11,100-10,700 14C yr BP, with renewed warming and wetting ~ 10,000
14C yr BP.
```

and then comments on the timing of this dry event in the tropics, relative to the Younger Dryas cold event

```
The similarity of these and other records indicates that this climate
oscillations was not only a global event, but also that there were no
significant time lags between tropical and temperate components of the climate
system during the last deglaciation.
```

Having read the most important sections of the article, we read the remaining text and figures in more detail if additional information is required for use the article to be used as a reference in our own manuscript (see ►Chap. 12). In that case, we cite the article in the text as

```
Roberts N, Taieb M, Barker P, Damnati B, Icole M, Williamson D (1993) Timing
of the Younger Dryas event in East Africa from lake-level changes. Nature
366:146-148
```

2

in the reference list of our article. The format used for references can vary significantly depending on the requirements of the particular journal, provided in their *Guidelines for Authors*. In the text itself, the article is cited as

```
Roberts et al. (1993) presented ...
```

where, as explained previously, et al. stands *et alii*, which is Latin for *and others*, since the article has more than two authors. If there are two authors only, e.g., N. Roberts and M. Taieb, the reference is cited as

```
Roberts and Taieb (1993) presented ...
```

or

```
Roberts & Taieb (1993) presented ...
```

In most cases, a reference occurs at the end of a statement rather than being incorporated in the text itself. The references then reads

```
East African climate was dry during the Younger Dryas (Roberts et al.,1993).
```

or in the case of two authors

```
East African climate was dry during the Younger Dryas (Roberts and Taieb,
1993).
```

or

```
East African climate was dry during the Younger Dryas (Roberts & Taieb,
1993).
```

depending on the format that each journal defines in its guidelines. In many cases more than one reference is cited after a scientific statement, such as

```
East African climate was dry during the Younger Dryas (Williamson et al.,
1993; Roberts et al., 1993).
```

with the references listed chronologically, with the oldest first. Bearing in mind the space limitations of most journals, however, we need to decide which references to actually cite for a statement, which brings us back to the original topic of this section. As an example, the maximum number of references in the main text in Nature and Science is thirty or fifty citations, depending on the type of paper, whereas there is no limit to the number of references in the electronic supplement.

2.5 Extracting Text, Data and Graphs from Literature

Having identified the most relevant article on a specific topic, we now wish to extract scientific information from that article, not only by reading and remembering that information, but also by taking text and figures from the electronic document for incorporation into a poster, a presentation slide or any other type of presentation, of course within the limits of copyright regulations. Most posters and presentations contain original research and the figures are therefore newly-designed by the researcher. In the bachelor's level course that inspired this book, however, the students were not conducting their own research and therefore prepared posters, abstracts, and presentations based on previously published work by other researchers. University lecturers, however, cover broader topics, usually complemented by their own original research.

The level of modifications that is possible for text and figures extracted from a published manuscript or book, prior to its incorporation into a presentation, depends on the way the document has been electronically saved. Computer graphics are generally stored and processed as either vector or raster data. Most of the data types encountered in the following chapter on processing geoscientific information are vector data, i.e., points, lines and polygons. Drainage networks, the outlines of geologic units, sampling locations and topographic contours are all examples of vector data. In ▶Chap. 6, coastlines are stored in a vector format while bathymetric and topographic data are saved in a raster format. Vector and raster data are often combined in a single data set, for example in order to display the course of a river on a satellite image. Raster data are often converted to vector data by digitizing points, lines or polygons. Conversely, vector data are sometimes transformed to raster data, which requires a degree of discretization and generalization. Images are generally presented as raster data, i.e., as a 2D array of color intensities. Images are everywhere in geosciences. Field geologists use aerial photographs and satellite images to identify lithologic units, tectonic structures, landslides, and other features within a study area. Petrologists and micropaleontologists use microscope images of minerals and fossils from either an optical microscope or a scanning electron microscope.

The PostScript (PS) format is used to store scalable vector graphics, raster images, and text. It was developed by John Warnock at PARC, the Xerox research institute. Warnock was also co-founder of Adobe Systems, where the EPS format was created. The PostScript vector format would have never become an industry standard without Apple Computers. In 1985 Apple needed a typesetter-quality controller for the new Apple LaserWriter printer and the Macintosh operating system, and adopted the PostScript format. The third partner in the history of PostScript was the Aldus company, the developer of the PageMaker software and now a part of Adobe Systems. The combination of Aldus PageMaker software, the PS format, and the Apple LaserWriter printer led to the creation of Desktop Publishing. The Encapsulated PostScript (EPS) format was then developed by Adobe Systems as a standard file format for importing and exporting PS files. Whereas a PS file is generally a single-page format containing either an illustration or a text, the purpose of an EPS file is to allow the inclusion of other pages, i.e., it is a file that can contain any combination of text, graphics and images. The Portable Document Format (PDF) designed by Adobe Systems in 1993 is now a true, self-contained, cross-platform document.

Most journal articles and electronic books are today available as PDF files, which may include editable text in a vector format, vector graphics, and raster images. The PDF file of the article by N. Roberts, however, is a scanned document in raster format, as is revealed by zooming in on the text and figures. It is therefore not possible to

2

extract text and figures in a vector format for further modification. Text recognition, automated vectorization, or manual digitizing can partly solve this problem, and pixel graphics could also be extracted without further editing and overlaid by vector additions in a separate layer (see ►Chap. 9).

More recent articles published in Nature, Science, or any other journal are available in a vector format. As an example, we use the article

```
Trauth MH, Maslin M, Deino MA, Deino A, Strecker MR (2005) Late Cenozoic
Moisture History of East Africa. Science 203:2051-2053.
```

published in 2005 in Science (►http://www.sciencemag.org). As with Nature, the Science webpage has a search field in which we can enter the title *Late Cenozoic Moisture History of East Africa* and quickly find the article.

```
Late Cenozoic Moisture History of East Africa
Martin H. Trauth, Mark A. Maslin, Alan Deino, and Manfred R. Strecker
Science 23 September 2005: 2051-2053. Published online 18 August 2005
[DOI:10.1126/science.1112964]
Abstract Full Text Full Text (PDF) Supporting Online Material
```

As well as the article being available as the abstract only, as the full text online, or as the full text in a PDF, it also includes supporting online material. Once again, the abstract is freely available:

```
Lake sediments in 10 Ethiopian, Kenyan, and Tanzanian rift basins suggest
that there were three humid periods at 2.7 to 2.5 million years ago
(Ma), 1.9 to 1.7 Ma, and 1.1 to 0.9 Ma, superimposed on the longer-term
aridification of East Africa. These humid periods correlate with increased
aridity in northwest and northeast Africa and with substantial global
climate transitions. These episodes could have had important impacts on the
speciation and dispersal of mammals and hominins, because a number of key
events, such as the origin of the genus Homo and the evolution of the
species Homo erectus, took place in this region during that time.
```

It provides the most important information and conclusions from the article. Access to the full text, whether online or as a PDF file, requires a subscription to the journal, or alternatively it can be accessed via a pay-per-view option. Having downloaded the PDF file, we quickly notice that it is in a true vector format since, for instance, we are able to mark and copy text fragments and to paste them into word processing software. Differences exist, however, in the way a PDF reader deals with the text, in particular with respect to indented text, special characters, and paragraph styles. Using Adobe Acrobat on a computer running the macOS operating system, for instance, right justification in the linefeeds of text is interpreted as a paragraph end marker at the end of each justified line, both in the abstracts and in the main text. For instance, the first paragraph of the main text appears as

```
Recent investigations of both terrestrial and
marine paleoclimate archives have led to a
concerted debate regarding the nature of Late
Cenozoic environmental changes in East Africa
and their influence on mammalian and hominin
evolution (1-3).
```

In this case, the paragraph end markers have to be removed either manually or using search and replace, prior to making further modifications of the text. Making the same experiment using Apple Preview avoids this problem, resulting in

```
Recent investigations of both terrestrial and marine paleoclimate archives
have led to a concerted debate regarding the nature of Late Cenozoic
environmental changes in East Africa and their influence on mammalian
and hominin evolution (1-3).
```

Another problem when extracting text fragments from an article is the incorrect transfer of subscripts, superscripts, and other paragraph styles. For instance, the term $^{40}Ar/^{39}Ar$ *age calibration* from the last sentence of the second paragraph of the main text becomes

```
Although much smaller than the lakes in the
western branch and often subaerially exposed,
these basins host a rich sedimentary record,
with intercalated volcaniclastic deposits that
permit high-precision 40Ar/39Ar age calibration
of lake-level highstands (5, 6) (Fig. 2).
```

in which the numbers *40* and *39* are transferred as a smaller font but not as superscripts, with Adobe Acrobat. Making the same experiment again using Apple Preview avoids this problem. Copying this sentence from the PDF viewed with Apple Preview again interprets the linefeeds correctly but puts hyphens and a space within words at linefeeds, such as *calibration* in the text below:

```
Although much smaller than the lakes in the western branch and often
subaerially exposed, these basins host a rich sedimentary record, with
intercalated volcaniclastic deposits that permit high-precision ⁴⁰Ar/³⁹Ar
age calibration of lake-level highstands (5, 6) (Fig. 2).
```

These errors again need to be corrected either manually or by using the search and replace option in text processing software.

In the next experiment we try to import the two figures included in the article by Trauth et al. (2005). We use the import feature of vector graphics software *Adobe Illustrator*. Opening the first page of the PDF file reveals that Figure 1, which shows a map of East Africa, is indeed in a vector format. After importing the figure we can click on individual elements such as polygons or text, and modify colors, line thicknesses and font sizes (see ► Chap. 8). However, importing the second page of the article, which includes Figure 2, reveals that this figure is in a pixel format, preventing any further modification. We suspect that the publisher may include data graphs, such as the compilation of lake records presented in Figure 2, in a pixel format in order to prohibit unwanted modification or manipulation of the author's original graphs. This figure, however, could be imported into pixel software such as *Adobe Photoshop* and modified with the tools available for editing photos and other pixel graphics (see ► Chap. 8). Furthermore, the pixel graphics can be imported into vector graphics software and overlaid with vector elements such as polygons and text (see ► Chap. 8).

2.6 Organizing Literature in a Computer

Familiarization with a new topic at the beginning of a research project requires the collection and reading of a large number of articles, books, and reports, and their organization as PDF files on a computer. To organize the collection of references and electronic documents, we need literature or bibliographic management software. A large number of both open source and commercial software products are available for managing your literature. We use the open source *BibDesk* (►http://bibdesk.source-forge.net) software for macOS as well as the commercial Clarivate Analytics *EndNote* (►http://www.endnote.com) and Labtiva's *Papers* (►http://papersapp.com), which are representative examples of the typical design of literature management software products.

Bibliographic management software typically consists of a searchable reference data base, often linked with online data bases such as Google Scholar and Clarivate Analytics Web of Science, and a file directory service for managing the PDF files on a hard drive. Most software packages can be integrated into word processing software such as *OpenOffice Writer*, *Microsoft Word* or *Apple Pages*, and reference lists are then generated automatically in a specific journal's layout from references included in the manuscript text. Alternatively, software based on the *BibTex* standard, such as *BibDesk* for the Unix-based macOS, or *KBibTex* for Linux or Unix, uses *LaTeX* to prepare formatted literature lists.

The most popular commercial software for organizing reference data bases is Clarivate Analytics' *EndNote* (►http://www.endnote.com). The software was first released in 1988 and is therefore one of the oldest reference management tools, designed before Adobe had released the PDF. The software was therefore primarily created to manage citation libraries rather than to organize PDFs on a hard disk. In the early days, the user had to type references into the data base manually, using an input mask with fields for authors, title, year and other kinds of information related to the article. Subsequently, most literature search data bases allowed users to export the search results in a format that could be directly imported into EndNote. Today, Endnote can also be used to organize PDF files. The software can be integrated with word processors and automatically generates reference lists in the format required by a specific journal. Most universities provide campus licenses that can be acquired at a reasonable price. EndNote is available for Microsoft Windows and Apple macOS operation systems.

Launching the EndNote X7 software brings up a single multi-panel user interface that includes a panel for browsing the library, a list of all references in the data base and a panel for previewing, searching and editing the references. Since EndNote is a product of Clarivate Analytics, it is linked to Clarivate Analytics Web of Science. We select *Web of Science Core Collection* from *Online Search...* from the Tools menu, select *Title* as the criterion for the search in the *Online Search* tab, type the keywords "*Younger Dryas*" "*East Africa*" in the search field, and then press *Search*. This yields a list of several references including the article by N. Roberts, which is the earliest article listed. Double-clicking the article in the list opens a new window displaying the data sheet for this particular reference with the categories *Author, Year, Journal, Volume, Issue, Pages*, and so forth. The data sheet also includes the abstract, number of citations, and the author's address. The *Reference* menu of the software includes a *Find full text* feature that automatically includes the PDF of the article if the user has an active subscription to Nature. Having included the reference and the file of the article in the data base, we can export the reference in a format of the user's choice for inclusion in a manuscript requiring a specific format. As an example, the reference to the article by N. Roberts can be previewed and exported in the Nature format:

1. Roberts, N. *et al.* Timing of the Younger Dryas Event in East-Africa from Lake-Level Changes. *Nature* **366**, 146-148, doi:DOI 10.1038/366146a0 (1993).

Switching to the format of Elsevier's Quaternary Science Reviews journal creates the reference

Roberts, N., Taieb, M., Barker, P., Damnati, B., Icole, M., Williamson, D., 1993. Timing of the Younger Dryas Event in East-Africa from Lake-Level Changes. Nature 366, 146-148.

The format for the Proceedings of the National Academy of Sciences of the United States of America (PNAS) is as follows:

1. Roberts N, et al. (1993) Timing of the Younger Dryas Event in East-Africa from Lake-Level Changes. Nature 366(6451):146-148.

The user interface and features of the free BibDesk software are very similar to those of EndNote. The open source software is a BibTeX front end that uses LaTeX to export formatted reference lists. BibDesk (►http://bibdesk.sourceforge.net) is only available for Apple macOS; it was first released in 2002 and can be downloaded as a separate product, without LaTeX. A recommended alternative is to download the software as part of the *MacTex* bundle (►http://www.tug.org/mactex), which includes the official standard distribution of *TeX* for Macs. Installing MacTeX allows formatted reference lists to be prepared without requiring a separate LaTeX editor. We can select *Web of Science* from the *Searches* menu, causing a search panel to open in which entering the keywords "*Younger Dryas*" "*East Africa*" yields no results. Since there are no further options available for the search fields, we need to consult the Web of Science help page for further instructions. We learn that we have to include the search terms manually and modify our search into *TS = Younger Dryas East Africa*, where *TS* stands for *Topic Search*. This again yields the papers by D. Williamson and N. Roberts as the earliest references on the list. The data sheet for the article by N. Roberts includes a link to the Web of Science webpage but no direct link to the PDF of the article on the webpage of the journal. The *Preferences* menu provides numerous BibTeX styles with which to format the reference list for export. Choosing *apalike* as an example format and then clicking the *TeX Preview* button from the user interface results in the formatted output

ROBERTS, N., TAIEB, M., BARKER, P., DAMNATI, B., ICOLE, M., and WILLIAMSON, D. (1993). Timing of the younger dryas event in east-africa from lake-level changes. *Nature*, 366(6451):146–148.

which follows the capitalized format used in the Web of Science data base for the authors' names. The commercial EndNote software clearly corrects this mistake and results in the output seen previously.

An award-winning commercial alternative to EndNote and BibDesk for Apple macOS and Microsoft Windows is Labtiva's *Papers* (►http://papersapp.com). The software was released in 2007 and is currently available as Version 3, which provides numerous additional features to those of the first version. In contrast to EndNote and its open source counterpart BibDesk, this software was designed from the start to maintain large libraries of PDF documents comprising journal articles, books, theses, and other types of scientific literature. As with the other tools, Papers offers repository searches using the Web of Science and other data bases, editing of the file metadata using data sheets, full-screen reading, and commenting, as well as multiple import and export features. A great advantage of the software is the very straight-forward approach to rename and reorganize the PDF files on a hard drive. As soon as a PDF file has been imported into the software, it is renamed in a way defined by the user. In our example, for instance, the PDF file of the article by N. Roberts is renamed *Nature 1993 Roberts.pdf* and copied into the *1993* folder and a *Roberts* subfolder in the *Papers* directory, in the user's home directory. This contrasts with the way EndNote organizes its Library as it first creates a directory named *EndNote Library.Data*, then a subdirectory named *Library Trauth.Data*, in which the file named *My EndNote Library.enl* is located. The directory *EndNote Library.Data* also contains a subdirectory, among others, named *PDF*, which in turn contains the folder with a numerical name, such as *3829176425*. This folder contains the PDF renamed as *Roberts-1993-Timing of the Younger Dryas Event.pdf*.

BibDesk, Endnote, and Papers are only three of the reference management tools that are available for different operating systems, but they are representative examples. All of these tools help in managing rapidly expanding reference data bases and PDF collections during the course of a typical earth sciences research project.

Recommended Reading

Hirsch JE (2005) An index to quantify an individual's scientific research output. Proc Natl Acad Sci 102:16569–16572

Langenberg H (2010) Editorial—Number crunch. Nat Geosci 3:445

Roberts N, Taieb M, Barker P, Damnati B, Icole M, Williamson D (1993) Timing of the Younger Dryas event in East Africa from lake-level changes. Nature 366:146–148

Trauth MH, Maslin M, Deino MA, Deino A, Strecker MR (2005) Late Cenozoic Moisture History of East Africa. Science 203:2051–2053

Williamson D, Taieb M, Damnati B, Icole M, Thouveny M (1993) Equatorial extension of the Younger Dryas event—rock magnetic evidence from Lake Magadi (Kenya). Global Planet Change 7:235–242

Internet Resources for Earth Science Data

Electronic Supplementary Material The online version of this chapter (https://doi.org/10.1007/978-3-662-56203-1_3) contains supplementary material, which is available to authorized users.

3

3.1 Introduction

This chapter deals with electronic data resources, searching for specific data, transferring data between servers and computers (*clients*), and data storage. ▶Section 3.2 introduces the systems available for data storage, starting with a historical retrospective and ending with the most recent storage systems such as DVDs, CDs, and flash memory. ▶Section 3.3 discusses the various data and file formats used. These include ASCII and binary file formats, generic binary file formats such as the JPEG format for images or the PS formats for vector graphics, and software-specific file formats (such as the Microsoft Word file DOC format or the MATLAB binary MAT format). ▶Section 3.4 explains how data are transferred from one computer to another via data networks: it provides overviews of the history of data transfer and of the established transfer techniques and protocols. The remaining sections in this chapter go on to present several examples of data resources and searches, which are then used in subsequent chapters for the visualization and presentation of geoscientific information (▶Sects. 3.5–3.8). ▶Section 3.9 concludes the chapter with methods for storing and organizing data in a computer.

3.2 Data Storage in a Computer

Humans first started using technology to store information outside their brains about 2.5 million years ago. At that time the first *stone tools* were made, and one human individual was therefore able to pass on the technique for making things without the use of words. There have been many developments since that time, including the evolution of language and, some time later, of writing. The love song of the princess of the Sumerian king Shu-Shin, dating from about 5,500 years before the present, is the earliest known writing in history. Early hieroglyphs were written on *clay* or *limestone tablets*, and later on *papyrus*, from which the Greek word βιβλος or *biblos* (meaning book), and the English word *Bible*, are derived. Bibles were for many years copied by ancient scribes, and these copies were then copied over and over again, including notes on the margins of the pages as well as many errors, until *typography* was invented in China in the eleventh century, and subsequently in Europe (by Johannes Gutenberg) in the mid-15th century. Copying books manually was then no longer necessary, nor were public readings of books in lecture halls.

Portable information has become increasingly compressed over the last two centuries through, for instance, the invention of microfilm in the mid-19th century, and the punched cards originally used for textile looms and fairground organs. Punched cards were used with computers until the mid-1980s, when they were replaced by magnetic tapes. Data on tapes are stored using a binary system, with the symbol *0* for *no hole* (using the terminology from punched cards) and *1* for *hole*. Tapes were then largely replaced by diskettes or *floppy disks*, invented by IBM and available in 8, 5¼ and 3½ in. sizes, making the winding of tapes no longer necessary. Of course the life span of these magnetic media (about 5–30 years) is much shorter than that of clay tablets (at least 7,500 years). On the other hand, there is no comparison between the 3.2 MB maximum storage capacity of a 3½ in. floppy disk (corresponding to about 11 million zeros or ones) and the very limited amount of space available on clay tablets.

The first *hard drive disks* were introduced by IBM in 1956 but the relationship between cost and storage capacity prevented them from being attractive for the mass market until the 1980s. The storage capacities of hard drive disks today can be is up to 3 TB. Medium-capacity *magneto-optical* (MO) *drives*, and the more affordable *ZIP drives* (100 MB, later 250 MB and even 750 MB), became very popular in the 1990s but were later replaced by the much cheaper rewritable *compact disks* (CDs). The CD was introduced by Sony and Philips Consumer Electronics in the late 1970s as an optical storage medium for music, and was restricted to a maximum storage capacity of 870 MB. The *Digital Versatile Disk* (DVD) was introduced by a consortium of Sony, Philips, Toshiba, and Time Warner in 1995 as a medium with greater storage capacity than a CD (from 4.7 to about 17 GB), but with similar dimensions. The lifespan of data stored on CDs and DVDs is at least 10 years, although manufacturers claim at least 30 years.

A *flash memory* on a computer storage chip, which can be electronically erased, was introduced by Toshiba in the 1980s and soon found its way into consumer products such as digital cameras, digital audio players, mobile phones, and more recently, even into laptops, as a very robust and fast storage medium. The maximum storage capacity of flash memory is today 32 GB but several chips can be bundled together in laptops and mobile phones to increase the total amount of memory available.

3.3 Data Formats in Earth Sciences

A computer generally stores data as *binary digits* or *bits*. A bit is analogous to a two-way switch with two states, on $= 1$ and off $= 0$. The bits are joined together to form larger groups, such as bytes consisting of 8 bits, in order to store more complex types of data. Such groups of bits are then used to encode data such as numbers or characters. Unfortunately, different computer systems and software use different schemes for encoding data. For instance, the characters in the widely-used text processing software Microsoft Word differ from those in Apple Pages. Exchanging binary data is therefore difficult between different computer platforms and software. Binary data can be stored in relatively small files if the researchers processing the files are using similar systems for data exchange. The transfer rate for binary data is generally faster than that for other file formats.

A number of different formats for exchanging data have been developed during recent decades. The classic example of a data format that can be used with different computer platforms and software is the *American Standard Code for Information Interchange* (ASCII), which was first published in 1963 by the American Standards Association (ASA). As a 7-bit code, ASCII consists of $2^7 = 128$ characters (codes 0–127). Whereas ASCII-1963 lacked lower-case letters, they were included in the ASCII-1967 update, together with various control characters, such as *escape* and *line feed*, and various symbols, such as brackets and mathematical operators. A number of variants have since appeared in order to facilitate the exchange of text written in non-English languages, such as the expanded ASCII, which contains 255 codes, e.g., the Latin-1 encoding.

In 1987 Microsoft Corporation introduced the cross-platform 8-bit *Rich Text Format* (RTF), which includes *escape sequences* allowing the text to be formatted using a *WYSIWYG* (*what you see is what you get*) type of text processor (►http://microsoft.com). Binary DOC files generated by *Microsoft Word* (recently replaced by DOCX files with XML support, where XML stands for Extensible Markup Language), ODT files generated by *OpenOffice Text*, and *pages* files generated by *Apple Pages*, contain much more formatting information than RTF files but at the expense of being less compatible. *Apple Pages*, however, contains converters for reading DOC files and converting them into *pages* files, but the latest 2016 release of *Microsoft Word* cannot read *pages* files. The same applies to exchanges between *Apple Numbers* files and files generated by *Microsoft Excel*, and between Apple Keynote files and files generated by *Microsoft PowerPoint*.

Similar incompatibilities exist between graphics software packages, such as between *Adobe Illustrator* and *Photoshop*, or *CorelDraw* and *Canvas* (as commercial products), and between *Inkscape* and *Gimp* (as open source products). Computer graphics are stored and processed as either vector data or raster data. Many different kinds of plots, such as line graphs or bar plots, and also more complex graphics such as those displaying drainage networks, the outlines of geologic units, sampling locations, or topographic contours, are all examples of vector data. Vector data use mathematical equations to describe geometrical primitives such as polygons and points. The use of these mathematical equations, basically describing the relationship between xy coordinates of polygons, makes it possible to modify the polygons in many ways, e.g., by editing the vertices, scaling the entire polygon, or by changing line thicknesses, colors or other properties of the polygon (see ►Chaps. 5 and 6). Images, however, are generally represented as raster data, i.e., as a 2D array of color intensities. The *raster data* are stored as 2D arrays of elements, each of which contains information on the color intensity values of the *pixels* (short for *picture elements*) (see ►Chap. 7).

Numerous formats are available for saving vector and raster data into a file, each of which has its own particular advantages and disadvantages. The choice of one format over another in an application depends on the way the images are to be used in a project, and whether or not the images are to be analyzed quantitatively. The *Compuserve Graphics Interchange Format* (GIF) was developed in 1987 for raster images using a fixed 8-bit colormap of $2^8 = 256$ colors. The GIF uses compression without any loss of data and was designed for fast transfer rates over the Internet. The limited number of colors means that it is not the right format for the smooth color transitions that occur in aerial photos or satellite images. It is, however, often used for line art, maps, cartoons and logos (►http://compuserve.com). The *Portable Network Graphics* (PNG), developed in 1994, was designed to replace GIF and supports 24-bit or 32-bit colors with lossless data compression.

The Microsoft Windows *Bitmap Format* (BMP) is the default image format for computers using the Microsoft Windows operating system. However, numerous converters also exist for reading and writing BMP files on other platforms. Various modifications of the BMP format are also available, some of them without compression and others with effective and rapid compression (►http://microsoft.com). The *Tagged Image File Format* (TIFF) was designed by Microsoft and the Aldus Corporation in 1986 and went on to become an industry standard for image-file exchange. A TIFF file includes an image file header, a directory, and the data, in all available graphics and image file formats. Some TIFF files even contain vector and raster versions of the same picture, as well as images at different resolutions and with different colormaps. The main advantage of TIFF files was originally their portability. A TIFF should perform on all computer platforms but, unfortunately, the numerous modifications of the TIFF that evolved in subsequent years often resulted in incompatibilities, to the extent that the TIFF is now often referred to as *Thousands of Incompatible File Formats*.

The *Joint Photographic Experts Group* (JPEG) was established in 1986 for the purpose of developing various standards for image compression. Although JPEG is an acronym for the committee name, it is now widely used as the name for an image compression and a file format released in 1992. This compression consists of grouping pixel values into 8 × 8 blocks and transforming each block using a discrete cosine transform. As a result, all unnecessary high-frequency information is deleted, which makes this form of compression irreversible. The advantage of the JPEG format is the availability of a three-channel, 24-bit, true color version. This allows images with smooth color transitions to be stored. The newer *JPEG-2000* format uses a wavelet transform instead of the cosine transform (►Sect. 5.8) (►http://jpeg.org).

The *PICT* format was developed by Apple Computers in 1984 as the default format for Macintosh graphics. It can be used for both raster images and vector graphics. PICT uses various methods for compressing data. The PICT 1 format only supports monochrome graphics, but PICT 2 supports a color depth of up to 32 bits. The PICT format is not supported on other platforms although some PC software tools can work with PICT files (►http://apple.com). The *PostScript* (PS) format and the *Encapsulated PostScript* (EPS) format have already been discussed in ►Chap. 2. An EPS file is designed to allow the inclusion of different PS files, thus allowing text, graphics, and images to be combined. The *Portable Document Format* (PDF) designed by Adobe Systems is now a truly self-contained cross-platform document. PDF files, also previously discussed in ►Chap. 2, contain the complete formatting of vector illustrations, raster images and text, or a combination of all these, including all necessary fonts (►http://adobe.com).

Most computer operating systems and software today allow the import and export of vector and raster graphics in most of these formats, although minor incompatibilities still exist, which can often be time consuming. Binary AI files generated by Adobe Illustrator or PSD files by Adobe Photoshop contain a far greater amount of editing information than other formats, but at the expense of being less compatible. In addition to these text, table, and graphics formats, various data compression and archive formats permit multiple files to be bundled together into a single archive file, and the file size to then be compressed. The ZIP format was introduced by Phil Katz in 1989 and is now used to compress data and software files for faster file transfer over the Internet. The TAR format is the corresponding method for file compression in the UNIX world, first introduced in 1988. Both ZIP and TAR are often used together to create GZIP files. Most computer operating systems contain software to compress and uncompress ZIP and TAR files, as well as other compressed files.

The next section discusses the transfer of data through the Internet, while subsequent sections illustrate the exchange and processing of geoscientific information in various data formats.

3

3.4 Data Transfer Between Computers

The Internet has dramatically improved scientific communication since the early 1990s. Its history extends back into the 1960s and early 1970s, when it was first established as the military communication network *Advanced Research Projects Agency Network* (ARPANET). Between the mid-1970s and the early 1980s, university networks were also connected, commissioned by the United States National Science Foundation (NSF) and then opened to other networks in 1988. The connection of individual networks was possible through the use of the standard Internet protocol suite *Transmission Control Protocol/Internet Protocol* (TCP/IP). The TCP/IP protocol defines the format of digital information transferred through a network and the rules for such transfers between computer hardware and networks, including authentication and error management. The use of the TCP/IP protocol since the 1970s, the commercialization of the network in the late 1980s, and the establishment of the World Wide Web (WWW) in the early 1990s, have facilitated in the enormous growth of the Internet, with over two billion users worldwide in 2011. A comprehensive historical summary of the Internet can be found at ▶http://en.wikipedia.org/wiki/Internet.

While the TCP/IP protocol is the main protocol for the Internet as a whole, it operates in concert with several other protocols that are used for communication between computers, for electronic mail services, for the transfer of large data sets, or for accessing the World Wide Web. One of the first network protocols was *Telnet*, which provided bidirectional communication between computers and servers using either a *text terminal* or console connection. The applications running a text terminal are command-line interpreters or shells, such as the *Command Window* in MATLAB, allowing commands or series of commands to be entered after a prompt, which are then executed after pressing *Enter* (see ▶Sect. 4.2). Examples of text terminal applications are *Terminal* and *Console* in UNIX-based operating systems such as Solaris, Linux, or macOS, and *cmd.exe* or *PowerShell* in Microsoft Windows. The most popular alternative is the *Secure Shell* (SSH) network protocol introduced in 1995, which sends information, in particular passwords, in an encrypted format.

In the early 1990s, before the introduction of specialized software for handling electronic mail, emails were handled using command-line tools or the text user-interface based *Pine* email software introduced by the University of Washington in 1989. The *Simple Mail Transfer Protocol* (SMTP) for email transport across the Internet was introduced in the early 1980s by the University of California, Berkeley, and its successor *Sendmail X* is still the standard protocol for emailing. Today, email client software such as *Microsoft Outlook* and Apple *Mail*, which are built into their respective operating systems, or the free and popular alternative Mozilla *Thunderbird* (▶http://mozilla.org/thunderbird), facilitate the management of email. The exchange of emails between servers and personal computers is managed either by the *Post Office Protocol* (POP, now in the POP3 version) or the more advanced *Internet Message Access Protocol* (IMAP).

The *File Transfer Protocol* (FTP) is used to transfer larger packages of data from one computer to another over a TCP-based network. The protocol was written in the early 1970s by Abhay Bhushan, who was also involved in designing the TCP/IP Internet architecture. Most servers also provide an SSH or *Secure File Transfer Protocol* (SFTP) using the SSH protocol for data encryption. Today, FTP and SFTP are often provided by Internet browsers when accessing larger files for download, with their use being obvious in the change from *http://* to *ftp://* in the software's address bar. The *http* in the address bar stands for the *Hypertext Transfer Protocol* (HTTP), which was introduced in 1991 for communicating on the *World Wide Web* (WWW). Hypertext is text with references (i.e., *hyperlinks*, or more simply, *links*) that allow the user to immediately access information related to the text, images, or any other type of information shown on a webpage. Webpages, including the server and client software (*browsers*), are written in the *HyperText Markup Language* (HTML) developed at the European Organization for Nuclear Research (CERN) in 1990 to provide and access webpages on the Internet. These webpages are identified by their *Uniform Resource Locator* (URL) in the address field of the browser. For example, the URL for the relevant article on Wikipedia is

```
http://en.wikipedia.org/wiki/Uniform_Resource_Locator
```

The World Wide Web today combines most of the Internet services described within this section, including mail, browsing remote directories, and file transfer. It also provides the ability to access and view multimedia information such as ebooks, music, and videos, as well as cloud and similar network services, thereby making a significant contribution to the accessibility of scientific information.

3.5 Internet Resources: When Was the Younger Dryas?

In ▶Chap. 2 we searched for the most relevant journal articles on the influence of the Younger Dryas in East Africa and found this paper by N. Roberts:

```
Roberts N, Taieb M, Barker P, Damnati B, Icole M, Williamson D (1993)
Timing of the Younger Dryas event in East Africa from lake-
level changes. Nature 366:146-148
```

We used the popular web-based, open source encyclopedia *Wikipedia* to learn that the Younger Dryas stadial was a geologically brief (1,200 year) cold-climate period between approximately 12,900 and 11,700 years ago (see ▶http://en.wikipedia.org/wiki/Younger_Dryas). The two papers cited as sources in the Wikipedia article were by Carlson (2010) and Alley (2000):

```
Carlson, A. E. (2013). "The Younger Dryas Climate Event" (PDF).
Encyclopedia of Quaternary Science. 3. Elsevier. pp. 126-34.

Alley, Richard B. (2000). "The Younger Dryas cold inter-
val as viewed from central Greenland". Quater-
nary Science Reviews. 19 (1): 213-226. doi:10.1016/
S0277-3791(99)00062-1.
```

Let us assume that we are working on a review article on the worldwide influence of the Younger Dryas and we wish to plot a time series from several different locations on Earth, covering the critical time interval. Ice cores retrieved from Greenland in the early 1990s have made a considerable contribution to our understanding of the exact onset, duration, and termination of the Younger Dryas cold interval. Let us obtain the

3

ice core data described in the article by R.B. Alley. We enter *Younger Dryas Ice Core* in the search field of

```
http://google.com
```

and obtain (in November 2016) a list of about 94,700 results. One of the first results on the list is

```
Younger Dryas - National Climatic Data Center
www.ncdc.noaa.gov/paleo/abrupt/data4.html
20.08.2008 - This near-glacial period is called the Younger Dryas,
named after a flower ... The ice core record at Dome C
(Figure 6) shows that climate ...
```

which mentions R.B. Alley's paper as the source of one of the important data sets related to the Younger Dryas, with a link to a webpage of the *National Centers for Environmental Information* (NCEI), formerly known as the *National Climatic Data Center* (NCDC), which is part of the *National Oceanic and Atmospheric Administration* (NOAA) (◻Fig. 3.1):

```
http://ncdc.noaa.gov/paleo/pubs/alley2000/alley2000.html
```

The webpage provides the title and author of the article, and the most important graph (showing the temperature and snow accumulation time series over the last 20,000 years from the GISP2 ice core), as well as the abstract, the cited references, and a link to the Elsevier *Science Direct* website. We click on *Data* to obtain the ice core data shown on the graph, which takes us to

```
ftp://ftp.ncdc.noaa.gov/pub/data/paleo/icecore/greenland/summit/gisp2/
isotopes/gisp2_temp_accum_alley2000.txt
```

Depending on which internet browser we are using, we may be required to log on as a *guest* before being directed to the text file. We notice that the protocol changes from HTTP to FTP, as indicated by the *Uniform Resource Locator* (URL) in the address field. The data webpage seems to be in an ASCII text format, also indicated by the name of the corresponding file containing the data *gisp2_temp_accum_alley2000.txt*. We save this file on the hard drive of our computer and open it in a text editor such as Apple *TextEdit*, Microsoft *WordPad*, *TextWrangler* (free software from ►http://barebones.com/products/textwrangler that allows line numbers to be shown), or the MATLAB *Editor* (which also provides line numbering). The first 70 lines of the file provide extensive information on the data set, the source, a description of the data, and the article to be cited if the data are used in another article or in a textbook. The file header also indicates that the file was updated a few years after publication of the corresponding article. Lines 75 to 1707 contain the temperature data in the second column (in degrees C) and the corresponding ages in the first column (in thousands of years before present), while lines 1717 to 3414 contain the snow accumulation data in the second column (in meters per year) and as before, the corresponding ages in the first column, from the Greenland ice core. Comparing the age columns for both data sets reveals differences in the ages for the temperature and the snow accumulation

WDC for Paleoclimatology

Home · Research · Data · Education · What's New · Features · Perspectives · Site Map

National Climatic Data Center, Asheville, North Carolina

The Younger Dryas cold interval as viewed from central Greenland

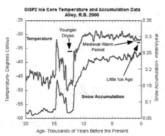

The Younger Dryas cold interval as viewed from central Greenland

Quaternary Science Reviews
Volume 19, Issues 1-5, 1 January 2000, Pages 213-226.

Richard B. Alley

Department of Geosciences and Environment Institute, The Pennsylvania State University, Deike Building, University Park, PA 16802, USA

ABSTRACT:
Greenland ice-core records provide an exceptionally clear picture of many aspects of abrupt climate changes, and particularly of those associated with the Younger Dryas event, as reviewed here. Well-preserved annual layers can be counted confidently, with only ~1% errors for the age of the end of the Younger Dryas ~11,500 years before present. Ice-flow corrections allow reconstruction of snow accumulation rates over tens of thousands of years with little additional uncertainty. Glaciochemical and particulate data record atmospheric-loading changes with little uncertainty introduced by changes in snow accumulation. Confident paleothermometry is provided by site-specific calibrations using ice-isotopic ratios, borehole temperatures, and gas-isotopic ratios. Near-simultaneous changes in ice-core paleoclimatic indicators of local, regional, and more-widespread climate conditions demonstrate that much of the Earth experienced abrupt climate changes synchronous with Greenland within thirty years or less. Post-Younger Dryas changes have not duplicated the size, extent and rapidity of these paleoclimatic changes.

To read or view the full study, please visit the Science Direct website.

◻ Fig. 3.1 Screenshot of the webpage describing the ice core data of Alley (2000), available online at the webpage of the National Climatic Data Center (NCDC), which is part of the National Oceanic and Atmospheric Administration (NOAA)

data points. We cannot therefore copy both data sets into a single table with a single column for the ages. Instead we mark, copy, and paste the two data sets separately into two files that we call *icecore_temperature_data.txt* and *icecore_snowaccumulation_data.txt*. The temperature data file *icecore_temperature_data.txt* includes one header line and 1632 lines of data:

```
     Age          Temperature (C)
 0.0951409          -31.5913
 0.10713            -31.622
 0.113149           -31.6026
 (cont'd)
```

The snow accumulation data file *icecore_snowaccumulation_data.txt* includes one header line and 1697 lines of data:

```
        Age            Accumulation
     0.144043          0.244106
     0.172852          0.246155
     0.20166           0.248822
     (cont'd)
```

3

In ►Chap. 5 we will use these data sets as examples to create plots with MATLAB. Alternatively, we could also use spreadsheet applications such as Microsoft *Excel*, Apple *Numbers* or the Open Office *Spreadsheet* to import, manipulate, and visualize the data. As an example, we will now demonstrate how to process the data in *icecore_ temperature_data.txt* using the Open Office Spreadsheet.

3.6 Internet Resources: Calibrating Radiocarbon Ages

This section deals with how to access an online database to calibrate radiocarbon ages. In ►Sect. 2.4 we learned from an article by N. Roberts that the Younger Dryas cold event resulted in a drier climate in tropical East Africa. Correlating the drier episode in tropical East Africa with the high latitude Younger Dryas event requires calibration of the ice core record from Alley (2000) against the lake-level record from Roberts et al. (1993). The Younger Dryas in Greenland lasted from 12.9 to 11.7 cal kyr BP (calibrated kiloyears before present, where the present is taken to be 1950). Figure 3 in Roberts et al. (1993) suggests that the level of Lake Magadi was low between 11.0 and 10.0 ^{14}C kyr BP (radiocarbon years before present). In order to compare the lake-level chronology from Roberts et al. (1993) with the ice core data from Alley (2000), we need to convert (calibrate) the radiocarbon ages into calibrated calendar ages (see Bradley 2015 for a detailed discussion of radiocarbon geochronologies).

Several online conversion tools are available for converting radiocarbon ages to calendar years. The most popular tool is the online version of the *CALIB 7* software by Stuiver and Reimer (1986) and Stuiver et al. (2005), using the IntCal13 and Marine13 radiocarbon age calibration curves covering the period from 0 to 50.000 yrs cal BP (□Fig. 3.2). The online documentation provides the required information on the various databases that are used to perform the calibration of the radiocarbon ages, depending on the environment (terrestrial or marine) from which the samples were taken. On the webpage

```
http://calib.org
```

we select *CALIB* and then access the online calibration tool by clicking on *Execute Version 7.1html*. We then get a menu with three options:

```
Data Input Menu
Calibration & Plot Options Menu
Marine Reservoir Correction Database (Lose input data)
```

We first select the *Calibration & Plot Options Menu* and enable the *1-sigma* and *2-sigma* errors, the *BP* age display, and the *Calibrated Curve*. We then select the *Data Input Menu* and type *11000* in the *Radiocarbon age BP* text field, *100* as the

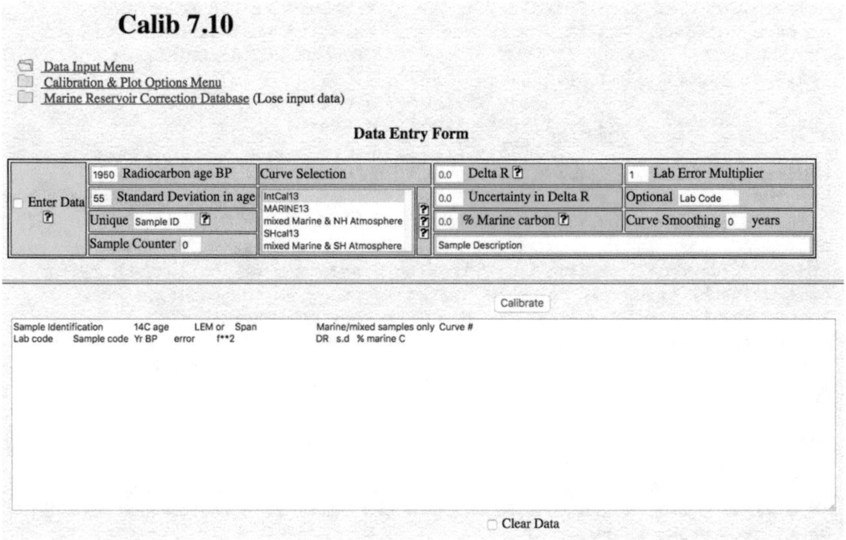

□ Fig. 3.2 Web interface of the online version of the *CALIB* software by Stuiver and Reimer (1986) and Stuiver et al. (2005), using the data sets of Reimer et al. (2013), hosted by the 14CHRONO Centre at Queens University, Belfast. The software helps to convert (calibrate) radiocarbon ages into calibrated calendar ages

Standard Deviation in age (this being the mean 1-sigma error reported in Table 1 of Roberts et al. 1993), *Onset YD* as the *Unique* sample identifier, *IntCal13* as the *Curve Selection* for terrestrial data, leave all other text input fields on default, and click *Enter Data*. We then change the *Unique* sample identifier to *Termination YD* and insert *10000* in the *Radiocarbon age BP* text field. Clicking *Calibrate* results in the following output:

```
            RADIOCARBON CALIBRATION PROGRAM*
                    CALIB REV7.1.0
        Copyright 1986-2016 M Stuiver and PJ Reimer
                *To be used in conjunction with:
    Stuiver, M., and Reimer, P.J., 1993, Radiocarbon, 35, 215-230.


Onset YD
Lab Code
Sample Description
Radiocarbon Age BP   11000 +/- 100
Calibration data set: intcal13.14c       # Reimer et al. 2013
  % area enclosed  cal BP age ranges      relative area under
                                          probability distribution
  68.3 (1 sigma)    cal BP 12763 - 12972  1.000
  95.4 (2 sigma)    cal BP 12718 - 13057  1.000
Median Probability: 12880


Termination YD
Lab Code
Sample Description
Radiocarbon Age BP   10000 +/- 100
```

3

```
Calibration data set: intcal13.14c       # Reimer et al. 2013
 % area enclosed  cal BP age ranges      relative area under
                                         probability distribution
  68.3 (1 sigma)   cal BP 11277 - 11627  0.946
                          11673 - 11697  0.054
  95.4 (2 sigma)   cal BP 11229 - 11833  0.966
                          11876 - 11947  0.034
Median Probability: 11520

References for calibration datasets:
Reimer PJ, Bard E, Bayliss A, Beck JW, Blackwell PG, Bronk Ramsey C,
 Buck CE, Cheng H, Edwards RL, Friedrich M, Grootes PM, Guilderson TP,
Haflidason H, Hajdas I, Hatté C, Heaton TJ, Hogg AG, Hughen KA,
Kaiser KF, Kromer B, Manning SW, Niu M, Reimer RW, Richards DA,
Scott EM, Southon JR, Turney CSM, van der Plicht J., IntCal13 and
MARINE13 radiocarbon age calibration curves 0-50000 years calBP
Radiocarbon 55(4). DOI: 10.2458/azu_js_rc.55.16947

Comments:
* This standard deviation (error) includes a lab error multiplier.
** 1 sigma = square root of (sample std. dev.^2  +  curve std. dev.^2)
** 2 sigma = 2 x square root of (sample std. dev.^2  +  curve std. dev.^2)
where ^2 = quantity squared.
[ ] = calibrated range impinges on end of calibration data set
0* represents a "negative" age BP
1955* or 1960* denote influence of nuclear testing C-14

NOTE: Cal ages and ranges are rounded to the nearest year which may be too
precise in many instances.  Users are advised to round results to the
nearest 10 yr for samples with standard deviation in the radiocarbon
age greater than 50 yr
```

The comprehensive information obtained from the online tool includes the calibrated ages for the onset of the lake-level lowstand in the Magadi basin

```
68.3 (1 sigma)   cal BP 12763 - 12972  1.000
95.4 (2 sigma)   cal BP 12718 - 13057  1.000
```

converted from 11.0 ^{14}C kyr BP, and the termination of this lowstand

```
68.3 (1 sigma)   cal BP 11277 - 11627  0.946
                        11673 - 11697  0.054
95.4 (2 sigma)   cal BP 11229 - 11833  0.966
                        11876 - 11947  0.034
```

converted from 10.0 ^{14}C kyr BP. The last column of the output corresponds to the mixing proportions of the two Gaussian distributions contributing to the age data. In our example, the value of 1.000 for the 12.763–12.972 cal kyr BP interval for the onset of the lowstand suggests that the arid episode started at about the same time as the high latitude Younger Dryas event (12.7 cal kyr BP). The termination of the Younger Dryas event (11.7 cal kyr BP) also falls close to the termination period for the lowstand of Lake Magadi, dated at 11.277–11.627 cal kyr BP (0.946 or 94.6% relative area under probability distribution), and suggests a synchronous climate in high and low latitudes (within the error bars).

3.7 Internet Resources: Insolation Data

Long-term climate fluctuations such as glacial-interglacial changes are explained by cyclic variations in the earth's orbit around the sun, known as Milankovitch cycles. These cycles affect the distribution of solar energy on our planet, causing changes in seasonality and in the heat transport between low and high latitudes, and hence in the climate. If we are running a project in tropical East Africa, we may wish to download details of these the orbital variations over the last 100 kyrs. The most popular place in which to search data of this type is again on the server of the National Centers for Environmental Information (NCEI), formerly known as NCDC of the NOAA.

```
https://www.ncei.noaa.gov
```

Selecting *Data Access* from the menu at the top of the webpage, then *Paleoclimatology* from the menu on the left, then *Paleoclimate Datasets*, and finally *Forcing*, takes us to the directory for the orbital variations data (◘ Fig. 3.3):

```
http://www.ncdc.noaa.gov/data-access/paleoclimatology-data/datasets/
climate-forcing
```

◘ **Fig. 3.3** Web interface of the *NOAA Paleoclimatology Program,* providing three data sets of orbital variations and insolation data. We use the data by Berger and Loutre (1991) as an example in this chapter (http://ncdc.noaa.gov/paleo/forcing.html)

As we can see, there are three different data sets of orbital variations:

```
Orbital Variations, 5,000,000 Years, Berger and Loutre 1991
Integrated Summer Insolation, 5,000,000 Years, Huybers 2006
Orbital Variations, -50 to +20 MYrs, Preliminary, Laskar et al. (from IMCCE)
```

The three teams have calculated the orbital variations independently and present slightly different results. Choosing the first data set, from A. Berger and M.F. Loutre, leads us to the FTP server of the NCDC.

```
ftp://ftp.ncdc.noaa.gov/pub/data/paleo/insolation/
```

Depending on which internet browser we are using, we may be required to log on as a *guest* before being directed to the text file. If your Internet browser does not launch an FTP client automatically you can use terminal software to access the site by typing

```
ftp ftp.ncdc.noaa.gov
```

after the prompt, logging on as *anonymous*

```
Trying 205.167.25.101...
Connected to ingest.ncdc.noaa.gov.
220 205.167.25.101 FTP server ready
Name (ftp.ncdc.noaa.gov:trauth): anonymous
```

and using your email address as the password.

```
331 Anonymous login ok, send your complete email address as your
password
Password:
230 Anonymous access granted, restrictions apply
Remote system type is UNIX.
Using binary mode to transfer files.
```

After logging on, you can change the directory to

```
ftp> cd pub/data/paleo/insolation/
```

and list the directory contents by typing

```
ftp> ls
```

which yields

```
150 Opening ASCII mode data connection for file list
-rw-rw-r--   1 2903 147864 Oct 13  1992 bein10.dat
-rw-rw-r--   1 2903 147864 Oct 13  1992 bein11.dat
-rw-rw-r--   1 2903 147864 Oct 13  1992 bein1.dat
-rw-rw-r--   1 2903 147864 Oct 13  1992 bein2.dat
-rw-rw-r--   1 2903 147864 Oct 13  1992 bein3.dat
-rw-rw-r--   1 2903 147864 Oct 13  1992 bein4.dat
-rw-rw-r--   1 2903 147864 Oct 13  1992 bein5.dat
-rw-rw-r--   1 2903 147864 Oct 13  1992 bein6.dat
-rw-rw-r--   1 2903 147864 Oct 13  1992 bein7.dat
-rw-rw-r--   1 2903 147864 Oct 13  1992 bein8.dat
-rw-rw-r--   1 2903 147864 Oct 13  1992 bein9.dat
```

```
-rw-rw-r--    1 2903     1443 Oct 13  1992 citation
-rw-rw-r--    1 2903      762 Oct 13  1992 contents.78
-rw-rw-r--    1 2903     1938 Oct 13  1992 contents.91
drwxrwxr-x    2 2903      512 Nov 14  2003 energy-balances
-rw-rw-r--    1 2903    63169 Oct 13  1992 insol91.dec
-rw-rw-r--    1 2903    63169 Oct 13  1992 insol91.jun
-rw-rw-r--    1 2903   375215 Oct 13  1992 orbit91
-rw-rw-r--    1 2903     3525 Mar 13  2008 readme_insolation.txt
226 Transfer complete
```

The directory contains a *readme* file that typically contains all relevant information on the data and the original references to be cited when using the data. We can download the file using

```
ftp> get readme_insolation.txt
```

The server's response is

```
local: readme_insolation.txt remote: readme_insolation.txt
229 Entering Extended Passive Mode (|||60776|)
150 Opening BINARY mode data connection for
readme_insolation.txt (3525 bytes)
100%
|***************************************************************
|  3525        29.74 MiB/s    00:00 ETA
226 Transfer complete
3525 bytes received in 00:00 (28.98 KiB/s)
```

and the ASCII text file *readme_insolation.txt* is delivered to our hard drive. We can open and view this file with a simple text editor. In the file, we can read that files named *bein1.dat* to *bein11.dat* each contain 100 kyrs of orbital calculations at 1 kyr intervals. The file *orbit91* contains the variations in the three orbital parameters eccentricity, obliquity, and precession, as well as mid-month insolation variations, for the following latitudes and months: 65 N (July), 65S (January), 15 N (July) and 15S (January). The files *insol91.dec* and *inso91.jun* contain December and June mid-month insolation variations at latitudes 90N, 60N, 30N, 0, 30S, 60S and 90S. We download the file *inso91.jun* in order to be able to plot the variations in June insolation on the 30N latitude, since this is believed to be the main driver for the African-Asian monsoonal circulation.

```
ftp> get insol91.jun
```

The file is again delivered to our hard drive but does not have a file extension. We change the file name from *inso91.jun* to *insol91jun.txt* to make it into a text file, which we can then open with a text editor:

```
                         INSOLATION
     90NJune 60NJune 30NJune  0 June 30SJune 60SJune 90SJune

  0   523.30  475.95  473.93  384.08  212.28   22.76    0.00
 -1   525.28  476.99  473.77  383.21  211.13   22.09    0.00
 -2   528.91  479.55  475.14  383.58  210.69   21.50    0.00
 -3   534.08  483.54  477.99  385.18  210.95   21.01    0.00
 (cont'd)
```

The first column shows the time in kyrs, using AD 1950 as the origin and negative numbers for the past, following the convention used for radiocarbon chronologies. All insolation values are in watts per square meter (W/m^2). We then close the FTP connection to the NCDC server by typing

```
ftp> quit
```

3

We can now easily import the data into MATLAB or any other software, in order to display it.

The second available data set was provided by Huybers (2006). Clicking the corresponding link on the NCDC webpage takes us to another webpage, which in turn provides an abstract and several links to individual data sets. Choosing the data description again leads us to the FTP server of the NCDC:

```
ftp://ftp.ncdc.noaa.gov/pub/data/paleo/climate_forcing/orbital_variations/
huybers2006insolation/huybers2006b.txt
```

There, the file *huybers2006b.txt* again provides all relevant information on the data and the references to be cited when using the data. In addition to providing numerous data files, this site also provides a MATLAB algorithm with which to calculate the insolation values provided on the server.

The third set of data is from J. Laskar and co-workers (Laskar et al. 1993, 2004). Clicking the corresponding link takes us to the server of the Institut de Mécanique Céleste et de Calcul des Éphémérides in Paris, France.

```
http://vo.imcce.fr/insola/earth/online/earth/earth.html
```

We then find a web directory by clicking on the *Source programs and data files here* link under the heading *Solutions La2004 from −50 Myr to +20 Myr*, which is what we are interested in.

```
Name                        Last modified      Size

INSOLN.LA2004.BTL.100.ASC  16-Mar-2004         3.7M
INSOLN.LA2004.BTL.250.ASC  26-Apr-2004         9.0M
INSOLN.LA2004.BTL.ASC      02-Mar-2004         4.4M
INSOLP.LA2004.BTL.ASC      02-Mar-2004         1.8M
Makefile                   18-Jan-2010         1.1K
README.TXT                 18-Jan-2010         8.0K
insola.f                   18-Jan-2010         12K
insola.par                 02-Mar-2004         198
insolsub.f                 18-Jan-2010         15K
prepinsol.f                02-Mar-2004         4.4K
prepsub.f                  02-Mar-2004         7.5K
```

We again find a readme file describing the data and listing the relevant references. We select the file *INSOLN.LA2004.BTL.ASC*, which includes the orbital parameters for the last 15 million years. According to the readme file, the first column of the table is time in kyrs, the second to fourth columns are the eccentricity, the obliquity (in radians) and the longitude of perihelion from moving equinox (in radians). The eccentricity data have recently been replaced by a new solution, which we can access via the *Data files here (revision 08 mars 2011)* link. We again rename the file as *INSOLN.LA2004BT-LASC.txt* and import it MATLAB or any other software to display the data.

3.8 Internet Resources: Tephrabase

The last example of Internet resources for geoscientific data is *Tephrabase*, which is hosted by the *School of GeoSciences* of the *University of Edinburgh*, United Kingdom (Newton 1996, Newton et al. 1997) (◘ Fig. 3.4):

```
http://tephrabase.org/
```

According to their webpage, Tephrabase was originally designed for use by scientists involved in tephrochronological research in north-west Europe. Tephrochronology uses volcanic ash layers as unique time markers within marine and terrestrial sedimentary sections. The ash layers can be either radiometrically dated, or correlated geochemically with other layers that have already been dated. Searches on Tephrabase are divided into tephra from either Iceland/Europe, Laacher See (a volcanic lake in Germany), or central Mexico. As an example, we can use Tephrabase to identify an as yet undated ash layer that we found in a marine sediment core from the North Atlantic. The oxygen isotope chronology of the core indicates an age for the ash within the range of 10–12 ^{14}C kyrs BP. This implies that the associated volcanic eruption occurred either during, or close to, Younger Dryas times.

Home   Background   Publications   Iceland/Europe   Laacher See   Central Mexico   Mapping/Tephrostratigraphy

Tephrochronology
Tephrochronology
TIQS2011
Acid Digestion

Recent Icelandic Eruptions
Bárðarbunga and Holuhraun 2014-15
Grímsvötn 2011
Eyjfjallajökull 2010

Links
Global Volcanism Program
INTAV
VHub.org
Alaskan Volcano Observatory
EarthChem
AntT
Global Volcano Model
VOGRIPA
OneGeology
INSPIRE, BGS
Institute of Earth Sciences, Iceland
volcano01010
Tephra Analytical Unit.

Tephrabase: A Tephrochronological Database

Tephra (volcanic ash) layers are now an invaluable tool in palaeoenvironmental studies, as well as a record of volcanic activity. The data produced by such research can be difficult to handle and disseminate. Tephrabase is a database of tephra layers found in Iceland, north-west and northern Europe, Russia and central Mexico. Details on the location, name, age and geochemistry of tephra layers are stored in the database, as well as information about relevant volcanoes and volcanic systems. A comprehensive reference database is also included. A new Laacher See supplementary data collection has been added.

Tephrabase has been redesigned. I hope that this makes it easier to get at the data you want to find. This is on an ongoing process and new features, including the automatic creation of tephrostratigraphic profiles have just been added and calculations of sediment accumulation rates for Iceland locations will be added shortly.

The data stored in Tephrabase can be broadly divided into five main categories. As Tephrabase develops these will change and it is hoped that connections can also be made to other online systems.

- Publications: Tephrabase has a frequently updated publications database (over 2600 entries) and all of the data in the database are linked to publications. Use this to search on keywords, authors or by year.
- Iceland/Europe: Tephrabase was originally developed as a source of information about Icelandic tephra layers and currently has details of over 2700 Icelandic-derived tephra layers found at over 300 sites in Iceland and Europe, as well as over 3000 major element analyses.
- Laacher See: This supplement to Tephrabase was developed with Felix Riede and contains details of over 430 sites where the c. 13 ka BP Laacher See tephra has been found and 16 locations with geochemical data.
- Central Mexico: Tephrabase also includes details of some 16 sites, with 53 tephra layers (over 700 geochemical analyses) in central Mexico.
- Mapping/Tephrostratigraphy: the locations of all sites in the database are shown on maps: sites in Europe

◘ **Fig. 3.4** The *Tephrabase* web interface, hosted by the School of GeoSciences of the University of Edinburgh, United Kingdom. The database helps with the identification of ash layers within the time range of 10–12 radiocarbon years before present

We select *Iceland/Europe* from the menu bar at the top of the webpage, and then click on *radiocarbon dated tephra layers* in the *Dates* section in order to perform a search based on ^{14}C dates. Please note that the dates provided on this webpage are radiocarbon ages and not calendar ages (see ▶Sect. 3.7). We select *between* and enter *10000* and *12000* ^{14}C kyrs BP, without thousand separators. The resulting page suggests the *Vedde Tephra* as the only possible ash layer within this time interval. Clicking on *Vedde Tephra* yields several locations in Sweden where the ash layer has been identified, geochemically analyzed, and dated at around 10.1–10.5 ^{14}C kyrs BP. Indeed, the Vedda Ash is an important time marker in deep-sea cores from the North Atlantic.

3.9 Organizing Data in a Computer

During the course of a project we accumulate large volumes of diverse data on the hard drives of our computers. In addition to a variety of information relating to the project, such as the addresses of collaborators in an address book, the project schedule in a calendar, and discussion threads with colleagues in email software, we also collect field photographs, measurements on samples, laboratory analyses, processed data, and published results. The rising flood of data acquired by the project requires well-organized data file directories on the computer.

We will, however, not discuss in this section specific software for data management or methods for analyzing large data sets. We will simply put together a few thoughts on how to store and organize the data in a computer using the on-board tools of the operating system. A filing system, of course, has to be adaptable to a growing project. As an example, we set up a top-level directory called *Younger Dryas East Africa*. We then create various levels of subfolders for various types of data, using the sorting function of the operating system to establish a hierarchy in our project directory, e.g., by labeling the subfolders 01, 02, 03 …, or by using years such as 2008, 2009, 2010, 2011 … to sort the subfolders.

```
Project Younger Dryas East Africa
    01 Research proposal and permits
        011 Outlines and ideas
        012 Materials
        013 Proposal text
        014 Proposal figures
        015 Correspondence
    02 Data from other people
    03 My field observations
        031 Field maps
        032 Field photos
        033 Field measurements
            0331 Method 1
            0332 Method 2
            ...
        034 Field samples
    04 My laboratory observations
        041 Laboratory photos
        042 Laboratory measurements
            0431 Method 1
            0432 Method 2
            ...
```

```
    043 Laboratory samples and subsamples
    044 Laboratory processed samples
05 My processed field and laboratory data
    051 Processing method 1
    052 Processing method 2
        . . .
06 Literature
07 Correspondence
08 Logistics
09 Publications
    091 Conference Presentations
    092 Manuscripts
        2008 Paper Younger Dryas East Africa
        2009 Paper Lake Magadi Lake Sediments
        2009 Paper Mag Sus Record Lake Magadi
        . . .
    093 Reports
        . . .
10 Other
```

When storing data on a computer, file names should not have any special characters or punctuation, other than full stop preceding the file extension. File extensions help to assign a file to a particular software. The most popular file extensions and their file types are

```
*.doc, *.xls, *.ppt       Microsoft Word, Excel and PowerPoint
*.docx, *.xlsx, *.pptx    dito, but Office Open XML based formats
*.txt, *.rtf              ASCII text and Microsoft Rich Text
*.htm, *.html             HyperText
*.exe, *.app              Executables under Windows and macOX
*.zip, *.tar              Compressed files
*.gif, *.jpg, *.bmp       Various raster graphics
*.ps, *.eps, *.pdf        Various vector graphics
*.ai, *.psd, *.indd       Adobe Illustrator, Photoshop and InDesign
*.m, *.mat                Matlab text and binary formats
```

Most of these file types have been described in the preceding text, or will be used in examples presented in the following chapters. The more specific file types include control characters that are not visible to the user but provide the relevant information concerning the formats of text and graphics, to the software. Most software products include converters for interpreting file types from other software products. In many examples, however, the appropriate file types for exchanging information are ASCII for text, TIFF, JPEG or PNG for pixel graphics, and PS, EPS or PDF for vector graphics. Other types of files will be explained in ▶Chaps. 5–7.

Meaningful file names should be used (e.g., *odp659_oxygenisotopes.xls* or *BG08_14C_Nov08.pdf*) containing information on the type of data stored in the file and including, for example, the name of a marine sediment core or sampling location, the method used to generate the data, and the date on which the file was created. Different versions of your files, such as processed data and manuscripts, should be retained in case you ever need to revert to a previous version of your work. If you are working on the same file as your collaborators, make sure that when your colleagues return the file they add their initials to the file name on the version that they have worked on, for example:

```
paper_youngerdryaseastafrica_vs1.doc
paper_youngerdryaseastafrica_vs2.doc
paper_youngerdryaseastafrica_vs2_MHT.doc
paper_youngerdryaseastafrica_vs3.doc
...
figure1_youngerdryaseastafrica_vs1.ai
figure1_youngerdryaseastafrica_vs2.ai
figure2_youngerdryaseastafrica_vs1.ai
figure2_youngerdryaseastafrica_vs1_MHT.ai
figure2_youngerdryaseastafrica_vs1_ES.ai
figure3_youngerdryaseastafrica_vs1.ai
...
```

Finally, be sure to back up your data regularly. Submissions of doctoral theses have in the past been delayed, either because a computer hard drive crashed a few days before the deadline, or because the entire computer system was stolen. Doctoral candidates are typically remiss about backing up during the last few weeks before submission and fully automated differential backups on a separate hard drive are therefore recommended. Examples of software that can be used to perform differential backups are *Apple Time Machine*, *Microsoft Windows Backup*, and the third-party *Genie-Soft Backup Manager*. Ideally, two backup hard drives should be used alternately on a daily basis and stored in different locations. In addition to these hard drive backup systems there are a number of online backup services available, such as *Dropbox* (►http://dropbox.com) and *Apple iCloud* (►http://apple.com/icloud).

Recommended Reading

Alley RB (2000) The younger dryas cold interval as viewed from central Greenland. Quat Sci Rev 19(1–5):213–226

Berger A, Loutre MF (1991) Insolation values for the climate of the last 10 million years. Quat Sci Rev 10:297–317

Bradley RS (2015) Paleoclimatology, reconstructing climates of the quaternary, 3rd edn. Academic Press, London

Carlson AE (2010) What caused the younger dryas cold event? Geology 38(4):383–384

Huybers P (2006) Early pleistocene glacial cycles and the integrated summer insolation forcing. Science 313:508–511

Laskar J, Robutel P, Joutel F, Gastineau M, Correia ACM, Levrard B (2004) A long-term numerical solution for the insolation quantities of the Earth. Astron Astrophys 428:261

Laskar J, Joutel F, Boudin F (1993) Orbital, precessional and insolation quantities for the Earth from −20 Myr to +10 Myr. Astron Astrophys 270:522

Newton AJ, Gittings B, Stuart N (1997) Designing a scientific database query server using the World Wide Web: the example of Tephrabase. Innovations in GIS, vol 4. Taylor and Francis, London, pp 251–266

Newton AJ (1996) Tephrabase. A tephrochronological database. Quat Newsl 78:8–13

Stuiver M, Reimer PJ (1986) A computer program for radiocarbon age calibration. Radiocarbon 28:1022–1030

Stuiver M, Reimer PJ, Reimer RW (2005) CALIB 5.0—WWW program and documentation (http://calib.qub.ac.uk)

Reimer PJ, Bard E, Bayliss A et al (2013) IntCal13 and Marine13 radiocarbon age calibration curves 0–50,000 years cal BP. Radiocarbon 55:1869–1887

Roberts N, Taieb M, Barker P, Damnati B, Icole M, Williamson D (1993) Timing of the younger dryas event in east africa from lake-level changes. Nat 366(6451):146–148

MATLAB as a Visualization Tool

Electronic Supplementary Material The online version of this chapter
(https://doi.org/10.1007/978-3-662-56203-1_4) contains supplementary material,
which is available to authorized users.

4.1 Introduction

MATLAB® is a software package developed by The MathWorks, Inc., founded by Cleve Moler, Jack Little and Steve Bangert in 1984, which has its headquarters in Natick, Massachusetts (▶http://www.mathworks.com). MATLAB was designed to perform mathematical calculations, to analyze and visualize data, and to facilitate the writing of new software programs. The advantage of this software is that it combines comprehensive math and graphics functions with a powerful high-level language. Since MATLAB contains a large library of ready-to-use routines for a wide range of applications, the user can solve technical computing problems much more quickly than with traditional programming languages, such as C++ and FORTRAN. The standard library of functions can be significantly expanded by add-on toolboxes, which are collections of functions for special purposes such as image processing, creating map displays, performing geospatial data analysis, or solving partial differential equations.

During the last few years MATLAB has become an increasingly popular tool in earth sciences. It has been used for finite element modeling, processing seismic data, analyzing satellite imagery, and generating digital elevation models from satellite data. The continuing popularity of the software is also apparent in published scientific literature, and many conference presentations have also made reference to MATLAB. Universities and research institutions have recognized the need for MATLAB training for staff and students, and many earth science departments across the world now offer MATLAB courses for undergraduates. The MathWorks, Inc. provides classroom kits for teachers at a reasonable price, and it is also possible for students to purchase a low-cost edition of the software. This student version provides an inexpensive way for students to improve their MATLAB skills.

The following sections contain a tutorial-style introduction to MATLAB, covering the setup on the computer (Sect. 4.2), the MATLAB syntax (Sects. 4.3 and 4.4), data input and output (Sects. 4.5 and 4.6), programming (Sects. 4.7 and 4.8), and visualization (Sect. 4.9). Advanced sections are also included on generating M-files to recreate graphics (Sect. 4.10), and on publishing M-files (Sect. 4.11). The reader is recommended to go through the entire chapter in order to obtain a good knowledge of the software before proceeding to the subsequent chapters of the book. A more detailed introduction can be found in the MATLAB Primer (MathWorks 2016), which is available in print form, online, and as a PDF file.

4.2 Getting Started with MATLAB

The software package comes with extensive documentation, tutorials and examples. The first three chapters of the book *MATLAB Primer* (MathWorks 2016) are directed at beginners. The chapters on programming, creating graphical user interfaces (GUIs) and development environments are aimed at more advanced users. Since *MATLAB Primer* provides all the information required to use the software, this introduction concentrates on the most relevant software components and tools used in the following chapters of this book.

After the installation of MATLAB, the software is launched either by clicking the shortcut icon on the desktop or by typing

```
matlab
```

in the operating system prompt. The software then comes up with several window panels (◧Fig. 4.1). The default desktop layout includes the *Current Folder* panel that lists the files in the directory currently being used. The *Command Window* presents the interface between the software and the user, i.e., it accepts MATLAB commands typed after the prompt >>. The *Workspace* panel lists the variables in the MATLAB workspace, which is empty when starting a new software session. In this book we mainly use the Command Window and the built-in *Editor*, which can be launched by typing

```
edit
```

By default, the software stores all of your MATLAB-related files in the startup folder named MATLAB. Alternatively, you can create a personal working directory in which to store your MATLAB-related files. You should then make this new directory the working directory using the *Current Folder* panel or the *Folder Browser* at the top of the MATLAB desktop. The software uses a *Search Path* to find MATLAB-related files, which are organized in directories on the hard disk. The default search path includes

◧ Fig. 4.1 Screenshot of the MATLAB default desktop layout including the *Current Folder* (left in the figure), the *Command Window* (center), and the *Workspace* (right) panels. This book uses only the *Command Window* and the built-in *Editor*, which can be called up by typing edit after the prompt. All information provided by the other panels can also be accessed through the Command Window

only the *MATLAB_R2016b* directory that has been created by the installer in the applications folder and the default working directory MATLAB. To see which directories are in the search path or to add new directories, select *Set Path* from the *Home* toolstrip of the *MATLAB* desktop, and use the *Set Path* dialog box. The modified search path is saved in the MATLAB preferences and the software will then in future direct MATLAB to use your custom path list.

4

4.3 The Syntax of MATLAB

The name MATLAB stands for *matrix laboratory*. The classic object handled by MATLAB is a *matrix*, i.e., a rectangular two-dimensional *array* of numbers. A simple 1-by-1 array is a *scalar*. Arrays with one column or row are *vectors*, time series, or other one-dimensional data fields. An *m*-by-*n* array can be used for a digital elevation model or a grayscale image. Red, green and blue (RGB) color images are usually stored as three-dimensional arrays, i.e., the colors red, green and blue are represented by an *m*-by-*n*-by-3 array.

Before proceeding, we need to clear the workspace by typing

```
clear
```

after the prompt in the Command Window. Clearing the workspace using `clear` is always recommended before working on a new MATLAB project to avoid name conflicts with previous projects. We can also go a step further, clear the content of the Command Window using `clc` and close all Figure Windows using `close all`. It is therefore recommended that a new MATLAB project should always start with the line

```
clear, clc, close all
```

Entering matrices or arrays in MATLAB is easy. To enter an arbitrary matrix, type

```
A = [2 4 3 7; 9 3 -1 2; 1 9 3 7; 6 6 3 -2]
```

which first defines a variable A, then lists the elements of the array in square brackets. The rows of A are separated by semicolons, whereas the elements of a row are separated by blank spaces, or alternatively, by commas. After pressing *return*, MATLAB displays the array

```
A =
    2    4    3    7
    9    3   -1    2
    1    9    3    7
    6    6    3   -2
```

Displaying the elements of A could be problematic for very large arrays such as digital elevation models consisting of thousands or millions of elements. To suppress the display of an array or the result of an operation in general, the line should end with a semicolon.

```
A = [2 4 3 7; 9 3 -1 2; 1 9 3 7; 6 6 3 -2];
```

The array A is now stored in the workspace and we can carry out some basic operations with it, such as computing the sum of elements,

```
sum(A)
```

which results in the display

```
ans =
    18    22    8    14
```

Since we did not specify an output variable, such as A for the array entered above, MATLAB uses a default variable ans, short for *answer* or *most recent answer*, to store the results of the calculation. In general, we should define variables since the next computation without a new variable name will overwrite the contents of ans.

The above example illustrates an important point about MATLAB: the software prefers to work with the columns of arrays. The four results of sum(A) are obviously the sums of the elements in each of the four columns of A. To sum all elements of A and store the result in a scalar b, we simply need to type

```
b = sum(sum(A));
```

which first sums the columns of the array and then the elements of the resulting vector. We now have two variables, A and b, stored in the workspace. We can easily check this by typing

```
whos
```

which is one of the most frequently-used MATLAB commands. The software then lists all variables in the workspace, together with information about their sizes or dimensions, number of bytes, classes and attributes (see ▸ Sect. 4.5 for details about classes and attributes of objects).

```
Name    Size        Bytes   Class     Attributes
A       4x4           128   double
ans     1x4            32   double
b       1x1             8   double
```

Note that by default MATLAB is case sensitive, i.e., A and a can define two different variables. In this context, it is recommended that capital letters be used for arrays that have two dimensions or more and lower-case letters for one-dimensional arrays (or vectors) and for scalars. However, it is also common to use variables with mixed large and small letters. This is particularly important when using descriptive variable names, i.e., variables whose names contain information concerning their meaning or purpose, such as the variable CatchmentSize, rather than a single-character variable a. We could now delete the contents of the variable ans by typing

```
clear ans
```

Next, we will learn how specific array elements can be accessed or exchanged. Typing

```
A(3,2)
```

simply yields the array element located in the third row and second column, which is 9. The array indexing therefore follows the rule (*row, column*). We can use this to replace single or multiple array elements. As an example we type

```
A(3,2) = 30
```

to replace the element A(3,2) by 30 and to display the entire array.

```
A =
     2      4      3      7
     9      3     -1      2
     1     30      3      7
     6      6      3     -2
```

If we wish to replace several elements at one time, we can use the *colon operator*. Typing

```
A(3,1:4) = [1 3 3 5]
```

or

```
A(3,:) = [1 3 3 5]
```

replaces all elements of the third row of the array A. The colon operator also has several other uses in MATLAB, for instance as a shortcut for entering array elements such as

```
c = 0 : 10
```

which creates a vector, or a one-dimensional array with a single row, containing all integers from 0 to 10. The resultant MATLAB response is

```
c =
     0   1   2   3   4   5   6   7   8   9   10
```

Note that this statement creates 11 elements, i.e., the integers from 1 to 10 and the zero. A common error when indexing arrays is to ignore the zero and therefore expect 10 elements instead of 11 in our example. We can check this from the output of whos.

```
Name      Size            Bytes  Class     Attributes
A         4x4               128  double
ans       1x1                 8  double
b         1x1                 8  double
c         1x11               88  double
```

The above command creates only integers, i.e., the interval between the array elements is one unit. However, an arbitrary interval can be defined, for example 0.5 units. This is later used to create evenly-spaced time vectors for time series analysis. Typing

```
c = 1 : 0.5 : 10
```

results in the display

```
c =
  Columns 1 through 6
    1.0000     1.5000     2.0000     2.5000     3.0000     3.5000
  Columns 7 through 12
    4.0000     4.5000     5.0000     5.5000     6.0000     6.5000
  Columns 13 through 18
    7.0000     7.5000     8.0000     8.5000     9.0000     9.5000
  Column 19
   10.0000
```

which autowraps the lines that are longer than the width of the Command Window. The display of the values of a variable can be interrupted by pressing *Ctrl+C* (*Control+C*) on the keyboard. This interruption affects only the output in the Command Window, whereas the actual command is processed before displaying the result.

MATLAB provides standard arithmetic operators for addition, +, and subtraction, -. The asterisk, *, denotes matrix multiplication involving inner products between rows and columns. For instance, we if multiply the matrix A with a new matrix B

```
B = [4 2 6 5; 7 8 5 6; 2 1 -8 -9; 3 1 2 3];
```

the matrix multiplication is then

```
C = A * B'
```

where ' is the complex conjugate transpose, which turns rows into columns and columns into rows. This generates the output

```
C =
   69    103    -79     37
   46     94     11     34
   53     76    -64     27
   44     93     12     24
```

In linear algebra, matrices are used to keep track of the coefficients of linear transformations. The multiplication of two matrices represents the combination of two linear transformations into a single transformation. Matrix multiplication is not commutative, i.e., A*B' and B*A' yield different results in most cases. Similarly, MATLAB allows matrix divisions representing different transformations, with / as the operator for right-hand matrix division and \ as the operator for left-hand division. Finally, the software also allows powers of matrices, ^.

In earth sciences, however, matrices are often simply used as two-dimensional arrays of numerical data rather than a matrix sensu stricto representing a linear transformation. Arithmetic operations on such arrays are carried out element-by-element. While this does not make any difference in addition and subtraction, it does affect multiplicative operations. MATLAB uses a dot, ., as part of the notation for these operations.

As an example multiplying A and B element-by-element is performed by typing

```
C = A .* B
```

which generates the output

```
C =
     8      8     18     35
    63     24     -5     12
     2      3    -24    -45
    18      6      6     -6
```

4.4 Array Manipulation

MATLAB provides a wide range of functions with which to manipulate arrays (or matrices). This section introduces the most important functions for array manipulation, which are used later in the book. We first clear the workspace and create two arrays, A and B, by typing

```
Clear, clc, close all

A = [2 4 3; 9 3 -1]
B = [1 9 3; 6 6 3]
```

which yields

```
A =
     2      4      3
     9      3     -1

B =
     1      9      3
     6      6      3
```

When we work with arrays we sometimes need to concatenate two or more arrays into a single array. We can either use `cat(dim,A,B)` with `dim=1` to concatenate the arrays A and B along the first dimension (i.e., along the rows), or use the function `vertcat` to concatenate the arrays A and B vertically. By typing either

```
C = cat(1,A,B)
```

or

```
C = vertcat(A,B)
```

we obtain (in both cases)

```
C =
     2      4      3
     9      3     -1
     1      9      3
     6      6      3
```

Similarly, we can concatenate arrays horizontally, i.e., concatenate the arrays along the second dimension (along the columns), either by typing

```
D = cat(2,A,B)
```

or by using the function `horzcat` instead

```
D = horzcat(A,B)
```

which both yield

```
D =
     2     4     3     1     9     3
     9     3    -1     6     6     3
```

When working with satellite images we often concatenate three spectral bands into three-dimensional arrays of the colors red, green and blue (RGB) (▶Sects. 4.5 and 7.4). We again use `cat(dim,A,B)`, but this time with `dim=3`, to concatenate the arrays A and B along the third dimension by typing

```
E = cat(3,A,B)
```

which yields

```
E(:,:,1) =
     2     4     3
     9     3    -1

E(:,:,2) =
     1     9     3
     6     6     3
```

Typing

```
whos
```

yields

```
Name      Size              Bytes  Class     Attributes
A         2x3                  48  double
B         2x3                  48  double
C         4x3                  96  double
D         2x6                  96  double
E         2x3x2                96  double
```

indicating that we have now created a three-dimensional array, as the size 2-by-3-by-2 suggests. Alternatively, we can use

```
size(E)
```

which yields

```
ans =
     2     3     2
```

to see that the array has 2 rows, 3 columns, and 2 layers in the third dimension. Using `length` instead of `size`,

```
length(A)
```

yields

```
ans =
    3
```

which tells us the dimension of the largest array only. Hence `length` is normally used to determine the length of a one-dimensional array (or vector), such the evenly-spaced time axis `c` that was created in ▶ Sect. 4.3.

MATLAB uses a matrix-style indexing of arrays, with the (1,1) element being located in the upper-left corner of arrays. Other types of data that are to be imported into MATLAB may follow a different indexing convention. As an example, digital terrain models (introduced in ▶ Sects. 7.3–7.5) often have a different way of indexing and therefore need to be flipped in an up-down direction or, in other words, about a horizontal axis. Alternatively, we can flip arrays in a left-right direction (i.e., about a vertical axis). We can do this by using `flipud` for flipping in an up-down direction and `fliplr` for flipping in a left-right direction

```
F = flipud(A)
F = fliplr(A)
```

yielding

```
F =
    9    3   -1
    2    4    3


F =
    3    4    2
   -1    3    9
```

In more complex examples we can use `circshift(A,K,dim)` to circularly shift (i.e., rotate) arrays by `K` positions along the dimension `dim`. As an example we can shift the array `A` by 1 position along the 2nd dimension (i.e., along the rows) by typing

```
G = circshift(A,1,2)
```

which yields

```
G =
    3    2    4
   -1    9    3
```

We can also use `reshape(A,[m n])` to completely reshape the array. The result is an `m`-by-`n` array `H` whose elements are taken column-wise from `A`. As an example we create a 3-by-2 array from `A` by typing

```
H = reshape(A,[3 2])
```

which yields

```
H =
    2    3
    9    3
    4   -1
```

Another important way to manipulate arrays is to sort their elements. As an example we can use `sort(C,dim,mode)` with `dim=1` and `mode='ascend'` to sort the elements of `C` in ascending order along the first array dimension (i.e., the rows). Typing

```
I = sort(C,1,'ascend')
```

yields

```
I =
     1     3    -1
     2     4     3
     6     6     3
     9     9     3
```

The function `sortrows(C,column)` with `column=2` sorts the rows of `C` according to the second column. Typing

```
J = sortrows(C,2)
```

yields

```
J =
     9     3    -1
     2     4     3
     6     6     3
     1     9     3
```

Array manipulation also includes the comparison of arrays, for example by checking whether elements in `A(i,j)` are also found in `B` using `ismember`. Typing

```
A, B

K = ismember(A,B)
```

yields

```
A =
     2     4     3
     9     3    -1

B =
     1     9     3
     6     6     3

K =
     0     0     1
     1     1     0
```

The `array` `L(i,j)` is zero if `A(i,j)` is not in `B`, and one if `A(i,j)` is in `B`. We can also locate elements within `A` for which a statement is true. For example we can locate elements with values less than zero and replace them with `NaNs` by typing

4

```
L = A;
L(find(L<0)) = NaN
```

or, more briefly

```
L(L<0) = NaN
```

which yields

```
L =
     2    4    3
     9    3  NaN
```

This is very useful when working with digital elevation models, in wich values below sea level are not relevant. Alternatively, we can replace data voids with values other than NaNs (such as −32768, for example), which are often used with digital terrain models (▶ Sects. 6.3–6.5). We can then determine which elements of an array are NaNs by typing

```
M = isnan(L)
```

which yields

```
M =
     0    0    0
     0    0    1
```

where NaNs are indicated by ones and non-NaN values are indicated by zeros. Which of the elements in array A are unique can be determined by typing

```
N = unique(A)
```

```
which yields
```

```
N =
    -1
     2
     3
     4
     9
```

The value of 3 occurs twice in A and the number of elements in N is therefore one less than in A.

4.5 Data Structures and Classes of Objects

The default data type or *class* in MATLAB is *double precision* or *double*, which stores data in a 64-bit array of floating-point numbers. Such floating-point numbers are approximations of real numbers that allow a maximum range of values in a limited numbers of bits. A double-precision array allows the sign of a number to be stored (bit 63), together with the exponent (bits 62 to 52), and roughly 16 significant decimal digits (bits 51 to 0). Typing

```
clear, clc, close all

realmin('double')
realmax('double')
```

yields the smallest and largest positive floating-point number in double precision

```
ans =
  2.2251e-308

ans =
  1.7977e+308
```

The actual number of floating point numbers is therefore limited by the number of bits available, in contrast to real numbers. The difference between 1.0 and the next largest double-precision number can be calculated using the floating-point relative accuracy eps by typing

```
eps(1.0)
```

which yields

```
ans =
  2.2204e-16
```

The round-off error depends on the value of the real number; it is, for example, different for `5.0`, as we can see by typing

```
eps(5.0)
```

which yields

```
ans =
  8.8818e-16
```

For real numbers there is, by definition, no such gap between consecutive numbers. The use of a finite number of floating-point numbers is limited by the number of available bits, due to the finite precision arithmetic of a computer. There are countless examples available with which to demonstrate this, but we will restrict ourselves to the simple example of the sine of π. Typing

```
sin(pi)
```

yields

```
ans =
  1.2246e-16
```

and not, as would be expected, zero. Since `pi` is only the nearest floating-point value to π, the sine of `pi` is not exactly zero but a value very close to zero.

Let us now look at some examples of arrays in order to familiarize ourselves with the different data types in MATLAB. For the first example we create a 3-by-4 array of random numbers with double precision by typing

```
clear

rng(0)
A = rand(3,4)
```

We use the function `rand` that generates uniformly distributed pseudorandom numbers within the open interval [0, 1]. To obtain identical data values, we use `rng(0)` to reset the random number generator by using the integer `0` as *seed*. Since we did not use a semicolon here we get the output

```
A =
    0.8147    0.9134    0.2785    0.9649
    0.9058    0.6324    0.5469    0.1576
    0.1270    0.0975    0.9575    0.9706
```

By default, the output is in a scaled fixed point format with 5 digits, e.g., `0.8147` for the `(1,1)` element of A. Typing

```
format long
```

switches to a fixed point format with 16 digits for double precision. Recalling A by typing

```
A
```

yields the output

```
A =
  Columns 1 through 2
   0.814723686393179    0.913375856139019
   0.905791937075619    0.632359246225410
   0.126986816293506    0.097540404999410

  Columns 3 through 4
   0.278498218867048    0.964888535199277
   0.546881519204984    0.157613081677548
   0.957506835434298    0.970592781760616
```

which autowraps those lines that are longer than the width of the Command Window. The command `format` does not affect how the computations are carried out, i.e., the precision of the computation results remains unchanged. The precision is, however, affected by converting the data type from *double* to 32-bit *single precision*. Typing

```
B = single(A)
```

yields

```
B =
    0.8147237    0.9133759    0.2784982    0.9648885
    0.9057919    0.6323593    0.5468815    0.1576131
    0.1269868    0.0975404    0.9575068    0.9705928
```

Although we have switched to `format long`, only 8 digits are displayed. The command `whos` lists the variables A and B with information on their sizes or dimensions, number of bytes, and classes

```
Name        Size             Bytes  Class      Attributes
A           3x4                 96  double
B           3x4                 48  single
```

The default class `double` is used in all MATLAB operations in which the physical memory of the computer is not a limiting factor, whereas `single` is used when working with large data sets. The double-precision variable A, whose size is 3-by-4 elements, requires $3 \times 4 \times 64 = 768$ bits or $768/8 = 96$ bytes of memory, whereas B requires only 48 bytes and so has half the memory requirement of A. Introducing at least one complex number to A doubles the memory requirement since both real and imaginary parts are double precision, by default. Switching back to `format short` and typing

```
format short
A(1,3) = 4i + 3
```

yields

```
A =
  Columns 1 through 2
    0.8147 + 0.0000i    0.9134 + 0.0000i
    0.9058 + 0.0000i    0.6324 + 0.0000i
    0.1270 + 0.0000i    0.0975 + 0.0000i

  Columns 3 through 4
    3.0000 + 4.0000i    0.9649 + 0.0000i
    0.5469 + 0.0000i    0.1576 + 0.0000i
    0.9575 + 0.0000i    0.9706 + 0.0000i
```

and the variable listing is now

```
Name        Size             Bytes  Class      Attributes
A           3x4                192  double     complex
B           3x4                 48  single
```

indicating the class `double` and the attribute `complex`.

MATLAB also works with even smaller data types such as 1-bit, 8-bit and 16-bit data, in order to save memory. These data types are used to store digital elevation models or images (see ▶Chaps. 6 and 7). For example m-by-n pixel RGB true color images are usually stored as three-dimensional arrays, i.e., the three colors are represented by an m-by-n-by-3 array (see ▶Chap. 7 for more details on RGB composites and true color images). Such multi-dimensional arrays can be generated by concatenating three two-dimensional arrays representing the m-by-n pixels of an image. First, we generate a 100-by-100 array of uniformly distributed random numbers in the range [0,1]. We then multiply the random numbers by 255 to get values between 0 and 255.

```
clear

rng(0)
I1 = 255 * rand(100,100);
I2 = 255 * rand(100,100);
I3 = 255 * rand(100,100);
```

The command `cat` concatenates the three two-dimensional arrays (8 bits each) into a three-dimensional array (3×8 bits $= 24$ bits).

```
I = cat(3,I1,I2,I3);
```

Since RGB images are represented by integer values between 0 and 255 for each color, we convert the 64-bit double-precision values to unsigned 8-bit integers using `uint8` (▶Sect. 8.2). The function `uint8` rounds the values in `I` to the nearest integer. Any values that are outside the range [0, 255] are assigned to the nearest endpoint (0 or 255).

```
I = uint8(I);
```

Typing `whos` then yields

```
Name          Size              Bytes  Class      Attributes
I             100x100x3         30000  uint8
I1            100x100           80000  double
I2            100x100           80000  double
I3            100x100           80000  double
```

Since 8 bits can be used to store 256 different values, this data type can be used to store integer values between 0 and 255, whereas using `int8` to create signed 8-bit integers generates values between -128 and $+127$. The value of zero requires one bit and there is therefore no space left in which to store $+128$. Finally, `imshow` can be used to display the three-dimensional array as a true color image.

```
imshow(I)
```

We next introduce *structure arrays* as a MATLAB data type. Structure arrays are multidimensional arrays with elements accessed by textual field designators. These arrays are data containers that are particularly helpful in storing any kind of information about a sample in a single variable. As an example we can generate a structure array `sample_1` that includes the image array `I` defined in the previous example as well as other types of information about a sample, such as the name of the sampling location, the date of sampling, and geochemical measurements, stored in a 10-by-10 array.

```
sample_1.location =  'Plougasnou';
sample_1.date = date;
sample_1.image = I;
sample_1.geochemistry = rand(10,10);
```

The first layer of the structure array `sample_1` contains a character array, i.e., a two-dimensional array of the data type `char` containing a character string. We can create such an array by typing

```
location = 'Plougasnou';
```

We can list the size, class and attributes of a single variable such as `location` by typing

```
whos location
```

and learn from

```
Name            Size            Bytes  Class      Attributes
Location        1x10               20  char
```

that the size of this character array `location` corresponds to the number of characters in the word *Plougasnou*. Character arrays are 16-bit arrays, i.e., $2^{16} = 65{,}536$ different characters can be stored in such arrays. The character string `location` therefore requires $10 \times 16 = 160$ bits or $160/8 = 20$ bytes of memory. In addition, the second layer datum in the structure array `sample_1` contains a character string generated by `date` that yields a string containing the current date in `dd-mm-yyyy` format. We access this particular layer in `sample_1` by typing

```
sample_1.date
```

which yields

```
ans =
    27-Jun-2014
```

as an example. The third layer of `sample_1` contains the image created in the previous example, while the fourth layer contains a 10-by-10 array of uniformly-distributed pseudorandom numbers. All layers of `sample_1` can be listed by typing

```
sample_1
```

resulting in the output

```
sample_1 =
        location: 'Plougasnou'
            date: '06-Oct-2009'
           image: [100x100x3 uint8]
    geochemistry: [10x10 double]
```

This represents a list of the layers `location`, `date`, `image` and `geochemistry` within the structure array `sample_1`. Some variables are listed in full, whereas larger data arrays are only represented by their size. In the list of the layers within the structure array `sample_1`, the array `image` is characterized by its size `100x100x3` and the class `uint8`. The variable `geochemistry` in the last layer of the structure array contains a 10-by-10 array of double-precision numbers. The command

```
whos sample_1
```

does not list the layers in `sample_1` but the name of the variable, the bytes and the class `struct` of the variable.

```
Name            Size            Bytes  Class      Attributes
sample_1        1x1             31546  struct
```

MATLAB also has *cell arrays* as an alternative to structure arrays. Both classes or data types are very similar and are containers of different types and sizes of data. The most important difference between the two is that the containers of a structure array are *named fields,* whereas a cell array uses *numerically-indexed cells.* Structure arrays are often used in applications where the organization of the data is particularly important. Cell arrays are often used when processing large data sets in count-controlled loops (▶ Sect. 4.7).

As an example of cell arrays we use the same data set that was used above to demonstrate structure arrays, with the layers of the structure array as the cells in the cell array. The cell array is created by enclosing the location name `Plougasnou`, the date, the image `I` and the 10-by-10 array of uniformly-distributed pseudorandom numbers in curly brackets.

```
C = {'Plougasnou' date I rand(10,10)}
```

Typing

```
C
```

lists the contents of the cell array

```
C =
  Columns 1 through 2
    'Plougasnou'    '27-Jun-2014'
  Columns 3 through 4
    [100x100x3 uint8]    [10x10 double]
```

which contains the location name and date. The image and the array of random numbers are too large to be displayed in the Command Window, but the dimensions and class of the data are displayed instead. We access a particular cell in `C`, e.g., the cell 2, by typing

```
C{2}
```

which yields

```
ans =
27-Jun-2014
```

We can also access the other cells of the cell array in a similar manner.

4.6 Data Storage and Handling

This section deals with how to store, import, and export data with MATLAB. Many of the data formats typically used in earth sciences have to be converted before being analyzed with MATLAB. Alternatively, the software provides several import routines to read many binary data formats in earth sciences, such as those used to store digital elevation models and satellite data. The simplest way to exchange data between a certain piece of software and MATLAB is using the ASCII format. Although the newer versions of MATLAB provide various import routines for file types such as Microsoft Excel binaries, most data arrive in the form of ASCII files. Consider a simple data set stored in a table such as

```
SampleID    Percent C       Percent S
101         0.3657          0.0636
102         0.2208          0.1135
103         0.5353          0.5191
104         0.5009          0.5216
105         0.5415          -999
106         0.501           -999
```

The first row contains the names of the variables and the columns provide the percentages of carbon and sulfur in each sample. The absurd value -999 indicates missing data in the data set. Two things have to be changed to convert this table into MATLAB format. First, MATLAB uses NaN as the representation for *Not-a-Number* that can be used to mark missing data or gaps. Second, a percent sign, %, should be added at the beginning of the first line. The percent sign is used to indicate non-executable text within the body of a program. This text is normally used to include comments in the code.

```
%SampleID   Percent C       Percent S
101         0.3657          0.0636
102         0.2208          0.1135
103         0.5353          0.5191
104         0.5009          0.5216
105         0.5415          NaN
106         0.501           NaN
```

MATLAB will ignore any text appearing after the percent sign and continue processing on the next line. After editing this table in a text editor, such as the MATLAB Editor, it can be saved as ASCII text file *geochem.txt* in the current working directory (◘Fig. 4.2). The MATLAB workspace should first be cleared by typing

```
clear, clc, close all
```

after the prompt in the Command Window. MATLAB can now import the data from this file with the load command.

```
load geochem.txt
```

MATLAB then loads the contents of the file and assigns the array to a variable geochem specified by the filename *geochem.txt*. Typing

```
whos
```

yields

```
Name        Size            Bytes   Class       Attributes
geochem     6x3               144   double
```

The command save now allows workspace variables to be stored in a binary format.

```
save geochem_new.mat
```

4

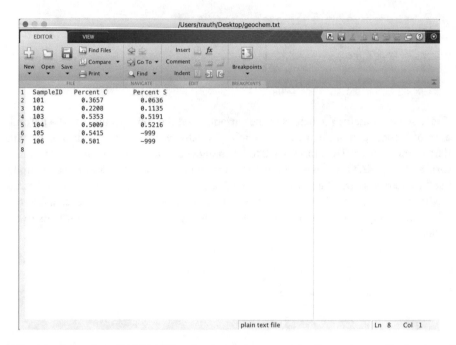

☐ Fig. 4.2 Screenshot of MATLAB *Editor* showing the content of the file *geochem.txt*. The first line of the text needs to be commented by a percent sign at the beginning of the line, followed by the actual data array. The -999 values need to be replaced by NaNs

MAT-files are double-precision binary files using *.mat* as extension. The advantage of these binary MAT-files is that they are independent of the computer platforms running different floating-point formats. The command

```
save geochem_new.mat   geochem
```

saves only the variable `geochem` instead of the entire workspace. The option `-ascii`, for example

```
save geochem_new.txt geochem -ascii
```

again saves the variable `geochem`, but in an ASCII file named *geochem_new.txt* in a floating-point format with 8 digits:

```
1.0100000e+02    3.6570000e-01    6.3600000e-02
1.0200000e+02    2.2080000e-01    1.1350000e-01
1.0300000e+02    5.3530000e-01    5.1910000e-01
1.0400000e+02    5.0090000e-01    5.2160000e-01
1.0500000e+02    5.4150000e-01           NaN
1.0600000e+02    5.0100000e-01           NaN
```

In contrast to the binary file *geochem_new.mat*, this ASCII file can be viewed and edited using the MATLAB Editor or any other text editor.

Such data files, especially those that are produced by electronic instruments, can look much more complicated than the example file *geochem.txt* with a single header line. In ▶Chaps. 6 and 7 we will read some of these complicated and extensive files, which are either binary or text files and usually have long headers describing the contents of the files. At this point, let us have a look at a variant of text files that contains not only one or more header lines but also unusual data types such as date and time, in a non-decimal format. We use the function textscan to perform this task. The MAT-LAB workspace should first be cleared by typing

```
clear
```

after the prompt in the Command Window. MATLAB can now import the data from the file *geochem.txt* using the textscan command.

```
fid = fopen('geochem.txt');
C = textscan(fid,'%u %f %f','Headerlines',1,'CollectOutput',1);
fclose(fid);
```

This script opens the file *geochem.txt* for *read only* access using fopen and defines the file identifier fid, which is then used to read the text from the file using textscan and to write it into the cell array C. The character string %u %f %f defines the conversion specifiers enclosed in single quotation marks, where %u stands for the 32-bit unsigned integer output class and %f stands for a 64-bit double-precision floating-point number. The parameter Headerlines is set to 1, which means that a single header line is ignored when reading the file. If the parameter CollectOutput is 1 (i.e., is true), textscan concatenates output cells with the same data type into a single array. The function fclose closes the file defined by fid. The array C is a cell array, which is a data type with indexed containers called cells (see ▶Sect. 4.5). The advantage of this data type is that it can store data of various types and sizes, such as character strings, double-precision numbers, and images in a single variable such as C. Typing

```
C
```

yields

```
C =
    [6x1 uint32]    [6x2 double]
```

indicating that C contains a 6-by-1 32-bit unsigned integer array, which is the sample ID, and a 6-by-1 double-precision array, which represents the percentages of carbon and sulfur in each sample. We can access the contents of the cells in C by typing

```
data1 = C{1}
data2 = C{2}
```

which yields

```
data1 =
    101
    102
    103
    104
    105
    106

data2 =
    0.3657    0.0636
    0.2208    0.1135
    0.5353    0.5191
    0.5009    0.5216
    0.5415       NaN
    0.5010       NaN
```

We now concatenate the two cells into one double-precision array `data`. We need first to change the class of `C{1}` into `double` or the class of the entire array `data` will be `uint32`. Typing

```
data(:,1) = double(C{1})
data(:,2:3) = C{2}
```

yields

```
101.0000    0.3657    0.0636
102.0000    0.2208    0.1135
103.0000    0.5353    0.5191
104.0000    0.5009    0.5216
105.0000    0.5415       NaN
106.0000    0.5010       NaN
```

The format of the data is as expected.

The next examples demonstrate how to read the file *geophys.txt,* which contains a single header line but also the date (in an *MM/DD/YY* format) and time (in an *HH:MM:SS.SS* format). We again use `textscan` to read the file,

```
clear

fid = fopen('geophys.txt');
data = textscan(fid,'%u %f %f %f %s %s','Headerlines',1);
fclose(fid);
```

where we skip the header, read the first column (the sample ID) as a 32-bit unsigned integer (`uint32`) with specifier `%u`, the next three columns X, Y, and Z as 64-bit double-precision floating-point numbers (`double`) with specifier `%f`, and then the date and time as character strings with specifier `%s`. We then convert the date and time to serial numbers, where a serial date number of 1 corresponds to *Jan-1-0000*. The year *0000* is merely a reference point and is not intended to be interpreted as a real year.

```
data_date_serial = datenum(data{5});
data_time_serial = datenum(data{6});
```

Finally, we can convert the date and time serial numbers into a data and time array by typing

```
data_date = datevec(data_date_serial)
data_time = datevec(data_time_serial)
```

which yields

```
data_date =
      2013    11    18    0    0    0
      2013    11    18    0    0    0
      2013    11    18    0    0    0
      2013    11    18    0    0    0
      2013    11    18    0    0    0

data_time =
   1.0e+03 *
   2.0130    0.0010    0.0010    0.0100    0.0230    0.0091
   2.0130    0.0010    0.0010    0.0100    0.0230    0.0102
   2.0130    0.0010    0.0010    0.0100    0.0230    0.0504
   2.0130    0.0010    0.0010    0.0100    0.0240    0.0051
   2.0130    0.0010    0.0010    0.0100    0.0240    0.0233
```

The first three columns of the array data_date contain the year, month and day. The fourth to sixth columns of the array data_time contain the hour, minute and second.

We can also write data to a formatted text file using fprintf. As an example we again load the data from *geochem.txt* after we have commented out the first line and replaced -999 with NaN. Instead of using load geochem.txt, we can type

```
clear

data = load('geochem.txt');
```

to load the contents of the text file into a double-precision array data. We write the data to a new text file *geochem_formatted.txt* using fprintf. Since the function fprintf writes all elements of the array data to the file in column order we need to transpose the data before we save it.

```
data = data';
```

We first open the file using the permission w for *writing*, and discard the existing contents. We then write data to this file using the formatting operators %u for unsigned integers and %6.4f for fixed-point numbers with a field width of six characters and four digits after the decimal point. The control character \n denotes a new line after each line of three numbers.

```
fid = open('geochem_formatted.txt','w');
fprintf(fid,'%u %6.4f %6.4f\n',data);
fclose(fid);
```

We can view the contents of the file by typing

```
edit geochem_formatted.txt
```

which opens the file *geochem_formatted.txt*

```
101    0.3657    0.0636
102    0.2208    0.1135
103    0.5353    0.5191
104    0.5009    0.5216
105    0.5415       NaN
106    0.5010       NaN
```

4

in the MATLAB Editor. The format of the data is as expected.

4.7 Control Flow

Control flow in computer science helps to control the order in which computer code is evaluated. The most important kinds of control flow statements are count-controlled loops such as `for` loops and conditional statements such as `if-then` constructs. Since in this book we do not deal with the programming capabilities of MATLAB in any depth, the following introduction to the basics of control flow is rather brief and omits certain important aspects of efficient programming, such as the pre-allocation of memory prior to using `for` loops, and instructions on how the use of `for` loops can be avoided by vectorization of the MATLAB code. This introduction is instead limited to the two most important kinds of control flow statements: the aforementioned `for` loops and the `if-then` constructs. Readers interested in MATLAB as a programming environment are advised to read the more detailed chapters on control flow in the MATLAB documentation (MathWorks 2014a and c).

The `for` loops, as the first example of a MATLAB language statement, execute a series of commands between `for` and `end` a specified number of times. As an example we use such a loop to multiply the elements of an array A by `10`, round the result to the nearest integer, and store the result in B.

```
clear

rng(0)
A = rand(10,1)
for i = 1 : 10
    B(i,1) = round(10 * A(i));
end
B
```

which yields

```
A =
    0.8147
    0.9058
    0.1270
    0.9134
    0.6324
    0.0975
    0.2785
    0.5469
    0.9575
    0.9649
```

```
B =
     8
     9
     1
     9
     6
     1
     3
     5
    10
    10
```

The result is as expected. We can expand the experiment by using a nested `for` loop to create a 2D array B.

```
rng(0)
A = rand(10,3)
for i = 1 : 10
    for j = 1 : 3
        B(i,j) = round(10 * A(i,j));
    end
end
B
```

which yields

```
A =
    0.8147    0.1576    0.6557
    0.9058    0.9706    0.0357
    0.1270    0.9572    0.8491
    0.9134    0.4854    0.9340
    0.6324    0.8003    0.6787
    0.0975    0.1419    0.7577
    0.2785    0.4218    0.7431
    0.5469    0.9157    0.3922
    0.9575    0.7922    0.6555
    0.9649    0.9595    0.1712


B =
     8     2     7
     9    10     0
     1    10     8
     9     5     9
     6     8     7
     1     1     8
     3     4     7
     5     9     4
    10     8     7
    10    10     2
```

This book tries to make all of the recipes independent of the actual dimensions of the data. This is achieved by the consistent use of `size` and `length` to determine the size of the data instead of using fixed numbers such as the 30 and 3 in the above example (▶ Sect. 4.4).

```
rng(0)
A = rand(10,3)
for i = 1 : size(A,1)
    for j = 1 : size(A,2)
        B(i,j) = round(10 * A(i,j));
    end
end
B
```

When working with larger data sets with many variables one might occasionally wish to automate array manipulations such as those described in ▶ Sect. 4.4. Let us assume, for example, that we want to replace all NaNs in all variables in the memory with -999. We first create a collection of four variables, each of which contains a single NaN.

```
clear

rng(0)
A = rand(3,3); A(2,1) = NaN
BC = rand(2,4); BC(2,2) = NaN
DE = rand(1,2); DE(1,1) = NaN
FG = rand(3,2); FG(2,2) = NaN
```

We list the variables in the workspace using whos and store this list in variables.

```
variables = who;
```

We then use a for loop to store the content of each variable in v using eval and then locate the NaNs in v using isnan (▶Sect. 4.4) and replace them with -999. The function eval executes a MATLAB expression stored in a text string. We assign the value of v to the variable in the base workspace and then clear the variables i, v and variables, which are no longer needed.

```
for i = 1 : size(variables,1)
    v = eval(variables{i});
    v(isnan(v)==1) = -999;
    assignin('base',variables{i},v);
    eval(variables{i})
end

clear i v variables
```

Comparing the variables before and after the replacement of the NaNs with -999 reveals that the script works well and that we have successfully manipulated our data.

The second important statements to control the flow of a script (apart from for loops) are if-then constructs, which evaluate an expression and then execute a group of instructions if the expression is true. As an example we compare the value of two scalars A and B.

```
clear

A = 1
B = 2
if A < B
    disp ('A is less than B')
end
```

which yields

```
A is less than B
```

The script first evaluates whether A is less than B and, if it is, displays the message A is less than B in the Command Window. We can expand the if-then construct by introducing else, which provides an alternative statement if the expression is not true.

```
A = 1
B = 2
if A < B
    disp('A is less than B')
else
    disp('A is not less than B')
end
```

which yields

```
A is less than B
```

Alternatively, we can use elseif to introduce a second expression to be evaluated.

```
A = 1
B = 2
if A < B
    disp('A is less than B')
elseif A >= B
    disp('A is not less than B')
end
```

The for loops and if-then constructs are extensively used in the following chapters of the book. For other aspects of programming, please refer to the MATLAB documentation (MathWorks 2016).

4.8 Scripts and Functions

MATLAB is a powerful programming language. All files containing MATLAB code use *.m* as an extension and are therefore called *M-files*. These files contain ASCII text and can be edited using a standard text editor. However, the built-in Editor color-highlights various syntax elements such as comments in green, keywords such as *if*, *for* and *end* in blue, and character strings in pink. This syntax highlighting facilitates MATLAB coding.

MATLAB uses two types of M-files: scripts and functions. Whereas scripts are a series of commands that operate on data in the workspace, functions are true algorithms with input and output variables. The advantages and disadvantages of both types of M-file will now be illustrated by an example. We first start the Editor by typing

```
edit
```

This opens a new window named *untitled*. Next, we generate a simple MATLAB script by typing a series of commands to calculate the average of the elements of a data array x.

4

```
[m,n] = size(x);
if m == 1
    m = n;
end
sum(x)/m
```

The first line of the `if-then` construct yields the dimensions of the variable x using the command `size`. In our example x should be either a column vector, i.e., an array with a single column and dimensions $(m,1)$, or a row vector, i.e. an array with a single row and dimensions $(1,n)$. The `if` statement evaluates a logical expression and executes a group of commands if this expression is true. The `end` keyword terminates the last group of commands. In the example the `if-then` construct picks either m or n depending on whether $m==1$ is false or true. The double equal sign `==` makes element by element comparisons between the variables (or numbers) to the left and right of the equal signs and returns an array of the same size, made up of elements set to logical 1 where the relationship is true and to logical 0 where it is not true. In our example $m==1$ returns 1 if m equals 1 and 0 if m equals any other value. The last line of the `if-then` construct computes the average by dividing the sum of elements by m or n. We do not use a semicolon here in order to allow the output of the result. We can now save our new M-file as *average.m* and type

```
clear

x = [3 6 2 -3 8];
```

in the Command Window to define an example array x. We then type

```
average
```

without the extension *.m* to run our script and obtain the average of the elements of the array x as output.

```
ans =
    3.2000
```

After typing

```
whos
```

we see that the workspace now contains

```
Name        Size            Bytes  Class       Attributes
ans         1x1                 8  double
m           1x1                 8  double
n           1x1                 8  double
x           1x5                40  double
```

The listed variables are the example array x and the outputs of the function `size`, m and n. The result of the operation is stored in the variable `ans`. Since the default variable `ans` might be overwritten during one of the succeeding operations, we need to define a different variable. Typing

```
a = average
```

however, results in the error message

```
??? Attempt to execute SCRIPT average as a function.
```

We can obviously not assign a variable to the output of a script. Moreover, all variables defined and used in the script appear in the workspace; in our example these are the variables m and n. Scripts contain sequences of commands that are applied to variables in the workspace. MATLAB functions, however, allow inputs and outputs to be defined. They do not automatically import variables from the workspace. To convert the above script into a function we need to introduce the following modifications (❐Fig. 4.3):

```
function y = average(x)
%AVERAGE     Average value.
%    AVERAGE(X) is the average of the elements in the array X.

% By Martin Trauth, June 27, 2014

[m,n] = size(x);
if m == 1
    m = n;
end
y = sum(x)/m;
```

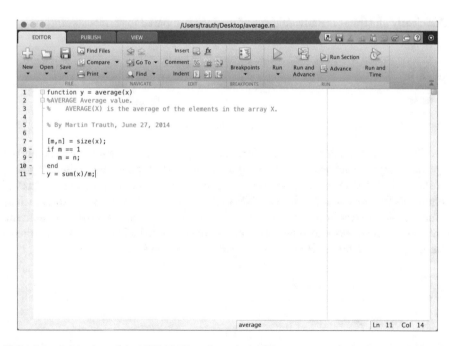

❐ **Fig. 4.3** Screenshot of the MATLAB *Editor* showing the function average. The function starts with a line containing the keyword function, the name of the function average, the input variable x, and the output variable y. The subsequent lines contain the output for help average, the copyright and version information, and also the actual MATLAB code for computing the average using this function

The first line now contains the keyword `function`, the function name `average`, the input `x` and the output `y`. The next two lines contain comments, as indicated by the percent sign, separated by an empty line. The second comment line contains the author's name and the version of the M-file. The rest of the file contains the actual operations. The last line now defines the value of the output variable `y`, and this line is terminated by a semicolon to suppress the display of the result in the Command Window. Next we type

```
help average
```

which displays the first block of contiguous comment lines. The first executable statement (or blank line in our example) effectively ends the help section and therefore the output of `help`. Now we are independent of the variable names used in our function. The workspace can now be cleared and a new data vector defined.

```
clear

data = [3 6 2 -3 8];
```

Our function can then be run by the statement

```
result = average(data);
```

This clearly illustrates the advantages of functions compared to scripts. Typing

```
whos
```

results in

```
Name           Size              Bytes  Class      Attributes
data           1x5                  40  double
result         1x1                   8  double
```

revealing that all variables used in the function do not appear in the workspace. Only the input and output as defined by the user are stored in the workspace. The M-files can therefore be applied to data as if they were real functions, whereas scripts contain sequences of commands that are applied to the variables in the workspace. If we want variables such as `m` and `n` to also appear in the memory they must be defined as *global* variables in both the function and the workspace, otherwise they are considered to be *local* variables. We therefore add one line to the function `average` with the command `global`:

```
function y = average(x)
%AVERAGE     Average value.
%    AVERAGE(X) is the average of the elements in the array X.

% By Martin Trauth, Nov 20, 2013

global m n
[m,n] = size(x);
if m == 1
    m = n;
end
y = sum (x)/m;
```

We now type

```
global m n
```

in the Command Window. After running the function as described in the previous example we find the two variables m and n in the workspace. We have therefore transferred the variables m and n between the function average and the workspace.

4.9 Basic Visualization Tools

MATLAB provides numerous routines for displaying data as graphics. This section introduces the most important graphics functions. The graphics can be modified, printed, or exported to be edited with graphics software other than MATLAB. The simplest function producing a graph of a variable y versus another variable x is plot. First, we define two one-dimensional arrays x and y, where y is the sine of x. The array x contains values between 0 and 2π with $\pi/10$ increments, whereas y is the element-by-element sine of x.

```
clear

x = 0 : pi/10 : 2*pi;
y = sin(x);
```

These two commands result in two one-dimensional arrays with 21 elements each, i.e., two 1-by-21 arrays. Since the two arrays x and y have the same length, we can use plot to produce a linear 2D graph of y against x.

```
plot(x,y)
```

This command opens a *Figure Window* named *Figure 1* with a gray background, an x-axis ranging from 0 to 7, a y-axis ranging from –1 to +1 and a blue line. We may wish to plot two different curves in a single plot, for example the sine and the cosine of x in different colors. The command

```
x = 0 : pi/10 : 2*pi;
y1 = sin(x);
y2 = cos(x);

plot(x,y1, '--',x,y2,'-')
```

creates a dashed blue line displaying the sine of *x* and a solid red line representing the cosine of this array (◼Fig. 4.4). If we create another plot, the window *Figure 1* will be cleared and a new graph displayed. The command figure, however, can be used to create a new figure object in a new window.

```
plot(x,y1,'--')
figure
plot(x,y2,'-')
```

4

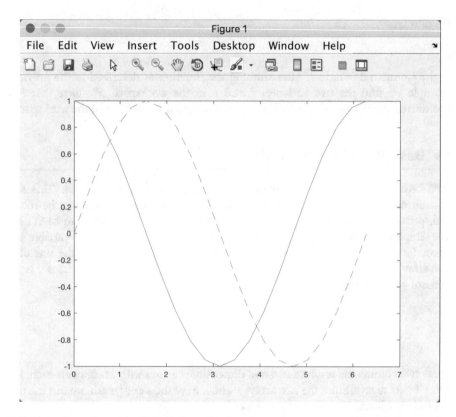

■ **Fig. 4.4** Screenshot of the MATLAB *Figure Window* showing two curves in different colors and line types. The Figure Window allows editing of all elements of the graph after selecting *Edit Plot* from the *Tools* menu. Double clicking on the graphics elements opens an options window for modifying the appearance of the graphics. The graphics can be exported using *Save as* from the *File* menu. The command *Generate Code* from the *File* menu creates MATLAB code from an edited graph

Instead of plotting both lines in one graph simultaneously, we can also plot the sine wave, hold the graph and then plot the second curve. The command hold is particularly important for displaying data while using different plot functions, for example if we wish to display the sine of x as a line plot and the cosine of x as a bar plot.

```
plot(x,y1,'r--')
hold on
bar(x,y2)
hold off
```

This command plots y1 versus x as a dashed red line using 'r--', whereas y2 versus x is shown as a group of blue vertical bars. Alternatively, we can plot both graphics in the same Figure Window but in different plots using the function subplot. The syntax subplot(m,n,p) divides the Figure Window into an *m*-by-*n* array of display regions and makes the *p*th display region active.

```
subplot(2,1,1), plot(x,y1,'r--')
subplot(2,1,2), bar(x,y2)
```

In this example the Figure Window is divided into two rows and one column. The 2D linear plot is displayed in the upper half of the Figure Window and the bar plot appears in the lower half. It is recommended that all Figure Windows be closed before proceeding to the next example. Subsequent plots would replace the graph in the lower display region only, or in other words, the last generated graph in a Figure Window. Alternatively, the command

```
clf
```

clears the current figure. This command can be used in larger MATLAB scripts after using the function `subplot` for multiple plots in a Figure Window.

An important modification to graphics is the scaling of the axis. By default, MATLAB uses axis limits close to the minima and maxima of the data. Using the command `axis`, however, allows the scale settings to be changed. The syntax for this command is simply `axis([xmin xmax ymin ymax])`. The command

```
plot(x,y1,'r--')
axis([0 pi -1 1])
```

sets the limits of the x-axis to 0 and π, whereas the limits of the y-axis are set to the default values -1 and $+1$. Important options of `axis` are

```
plot(x,y1,'r--')
axis square
```

which makes the x-axis and y-axis the same length, and

```
plot(x,y1,'r--')
axis equal
```

which makes the individual tick mark increments on the x-axis and y-axis the same length. The function `grid` adds a grid to the current plot, whereas the functions `title`, `xlabel` and `ylabel` allow a title to be defined and labels to be applied to the x-axis and y-axis.

```
plot(x,y1,'r--')
title('My first plot')
xlabel('x-axis')
ylabel('y-axis')
grid
```

These are a few examples how MATLAB functions can be used to edit the plot in the Command Window. More graphics functions will be introduced in the following chapters of this book.

4.10 Generating M-Files to Regenerate Graphs

MATLAB supports various ways of editing all objects in a graph interactively using a computer mouse. The *Edit Plot* mode of the Figure Window first needs to be activated by clicking on the arrow icon or by selecting *Edit Plot* from the *Tools* menu. The Figure

Window also contains some other options such as *Rotate 3D*, *Zoom* or *Insert Legend*. The various objects in a graph, however, are selected by double-clicking on the specific component, which opens the *Property Editor*. The Property Editor allows changes to be made to many features (or properties) of the graph such as axes, lines, patches and text objects.

The *Generate Code* option enables us to automatically generate the MATLAB code of a figure to recreate a similar graph with different data. We use a simple plot to illustrate the use of the Property Editor and the Generate Code option to recreate a graph.

4

```
clear, clc, close all

x = 0 : pi/10 : 2*pi;
y1 = sin(x);
plot(x,y1)
```

The default layout of the graph is that of ▪Fig. 4.4. Clicking on the arrow icon in the *Figure Toolbar* enables the *Edit Plot* mode. The selection handles of the graph appear, identifying the objects that are activated. Double-clicking an object in a graph opens the *Property Editor*.

As an example we can use the Property Editor to change various properties of the graph. Double-clicking the gray background of the Figure Window gives access to properties such as Figure Name, the Colormap used in the figure, and the Figure Color. We can change this color to light blue, represented by the light blue square in the 4th row and 8rd column of the color chart. Moving the mouse over this square displays the RGB color code [0.68 0.92 1] (see ▶Chap. 7 for more details on RGB colors). Activating the blue line in the graph allows us to change the line thickness to 2.0 and select a 15-point square marker. We can deactivate the Edit Plot mode of the Figure Window by clicking on the arrow icon in the Figure Toolbar.

After having made all necessary changes to the graph, the corresponding commands can even be exported by selecting Generate Code from the File menu of the Figure Window. The generated code displays in the MATLAB Editor.

```
function createfigure(X1, Y1)
%CREATEFIGURE(X1, Y1)
%   X1:  vector of x data
%   Y1:  vector of y data

%   Auto-generated by MATLAB on 27-Jun-2014 13:28:13

% Create figure
figure1 = figure('Color',[0.68 0.92 1]);

% Create axes
axes1 = axes('Parent',figure1,'ColorOrderIndex',2);
box(axes1,'on');
hold(axes1,'on');

% Create plot
plot(X1,Y1,'MarkerSize',15,'Marker','square','LineWidth',2);
```

We can then rename the function createfigure to mygraph and save the file as mygraph.m.

```
function mygraph(X1, Y1)
%MYGRAPH(X1,Y1)
%  X1:  vector of x data
%  Y1:  vector of y data
(cont'd)
```

The automatically-generated graphics function illustrates how graphics are organized in MATLAB. The function figure first opens a Figure Window. Using axes then establishes a coordinate system, and using plot draws the actual line object. The Figure section in the function reminds us that the light-blue background color of the Figure Window is represented by the RGB color coding [0.68 0.92 1]. The Plot section reveals the square marker symbol used and the line width of 2 points.

The newly-created function mygraph can now be used to plot a different data set. We use the above example and

```
clear

x = 0 : pi/10 : 2*pi;
y2 = cos(x);
mygraph(x,y2)
```

The figure shows a new plot with the same layout as the previous plot. The *Generate Code* function of MATLAB can therefore be used to create templates for graphics that can be used to generate plots of multiple data sets using the same layout.

Even though MATLAB provides abundant editing facilities and the *Generate Code* function even allows the generation of complex templates for graphics, a more practical way to modify a graph for presentations or publications is to export the figure and import it into a different software such as CorelDraw or Adobe Illustrator. MATLAB graphics are exported by selecting the command *Save as* from the File menu or by using the command print. This function exports the graphics, either as a raster image (e.g., JPEG or GIF) or as a vector file (e.g., EPS or PDF), into the working directory (see ►Chap. 7 for more details on graphic file formats). In practice, the user should check the various combinations of export file formats and the graphics software used for final editing of the graphics.

4.11 Publishing M-Files

Another useful feature of the software is the option to publish reports on MATLAB projects in various file formats such as HTML, XML, LaTeX and many others. This feature enables you to share your results with colleagues who may or may not have the MATLAB software. The published code includes formatted commentary on the code, the actual MATLAB code, and all results of running the code including the output to the Command Window and all graphics created or modified by the code. To illustrate the use of the publishing feature we create a simple example of a commented

MATLAB code to compute the sine and cosine of a time vector and display the results as two separate figures.

We start the Editor by typing `edit` in the Command Window, which opens a new window named `untitled`. An M-file to be published starts with a document title at the top of the file, followed by some comments that describe the contents and the version of the script. The subsequent contents of the file include sections of MATLAB code and comments, separated by the double percent signs %%. Whereas single percent signs % are known (from ►Sect. 4.8) to initiate comments in MATLAB, we now use double percent signs %% that indicate the start of new code sections in the Editor. The *code sections* feature, previously also known as *code cells* or *cell mode*, is a feature in MATLAB that enables you to evaluate blocks of commands called *sections* by using the buttons *Run, Run and Advance, Run Section, Advance,* and *Run and Time* on the *Editor Toolstrip* to evaluate either the entire script or parts of the script.

```
%% Example for Publishing M-Files
% This M-file illustrates the use of the publishing
% feature of MATLAB.
% By Martin Trauth, June 27, 2014

%% Sine Wave
% We define a time vector t and compute the sine y1 of t.
% The results are displayed as linear 2D graph y1 against x.
x = 0 : pi/10 : 2*pi;
y1 = sin(x);
plot(x,y1)
title('My first plot')
xlabel('x-axis')
ylabel('y-axis')

%% Cosine Wave
% Now we compute the cosine y2 of the same time vector and
% display the results.
y2 = sin(x);
plot(x,y2)
title('My first plot')
xlabel('x-axis')
ylabel('y-axis')

%%
% The last comment is separated by the double percent sign
% without text. This creates a comment in a separate cell
% without a subheader.
```

We save the M-file as *myproject.m* and click the *Publish* button in the *Publish Toolstrip*. The entire script is now evaluated and the Figure Windows pop up while the script is running. Finally, a window opens up that shows the contents of the published M-file. The document title and subheaders are shown in a red font whereas the comments are in black fonts. The file includes a list of contents with jump links to proceed to the chapters of the file. The MATLAB commands are displayed on gray backgrounds but the graphics are embedded in the file without the gray default background of Figure Windows. The resulting HTML file can be easily included on a course or project webpage. Alternatively, the HTML file and included graphics can be saved as a PDF file and shared with students or colleagues.

Recommended Reading

Etter DM, Kuncicky DC, Moore H (2014) Introduction to MATLAB. Prentice Hall, New Jersey

Gilat A (2010) MATLAB: an introduction with applications. Wiley, New York

Hanselman DC, Littlefield BL (2012) Mastering MATLAB 8. Prentice Hall, New Jersey

MathWorks (2016) MATLAB primer. The MathWorks Inc, Natick, MA

Palm WJ (2010) Introduction to MATLAB 7 for engineers. McGraw-Hill, New York

Quarteroni A, Saleri F, Gervasio P (2014) Scientific computing with MATLAB and Octave, 4th edn. Springer, Berlin

Trauth MH (2015) MATLAB® Recipes for Earth Sciences–4th Edition. Springer, Berlin

Visualizing 2D Data in Earth Sciences

Electronic Supplementary Material The online version of this chapter
(https://doi.org/10.1007/978-3-662-56203-1_5) contains supplementary material,
which is available to authorized users.

© Springer-Verlag GmbH Germany, part of Springer Nature 2018
M. H. Trauth and E. Sillmann, *Collecting, Processing and Presenting Geoscientific*
Information, Springer Textbooks in Earth Sciences, Geography and Environment,
https://doi.org/10.1007/978-3-662-56203-1_5

5.1 Introduction

In this chapter we demonstrate advanced two-dimensional visualization techniques in the form of graphical displays of the types of data typically encountered in earth sciences, using MATLAB. The first example displays graphically a temperature and snow accumulation time series for the last 20,000 years from the GISP2 ice core data presented by Alley (2000), in a single plot with an x-axis and two y-axes (►Sect. 5.2). ►Section 5.3 introduces the use of bar plots for displaying histograms, in which the temperature data used in the previous example is divided over equally spaced temperature intervals (called bins, or classes) and the counts per bin (the number of data points that fall within each bin) are displayed as a bar plot. The same frequency distribution for temperature values is then displayed as two- and three-dimensional pie charts, and the use of transparency is introduced (►Sect. 5.4). The visualization of directional data, which is particularly important in earth sciences, is illustrated in ►Sect. 5.5 although MATLAB is certainly not the best software for displaying data of this type. The last two sections of the chapter then introduce two specialized MATLAB scripts for displaying multiproxy data, such as pollen or microfossil records, and for visualizing stratigraphic logs of sedimentary sequences.

5.2 Line Graphs: Plotting Time Series in Earth Sciences

Our first example demonstrates how to create line graphs with MATLAB. After clearing the workspace and the Command Window and closing all Figure Windows, we load the GISP2 data for temperature and snow accumulation variations (from Alley 2000) from the two files created in ►Chap. 3.

```
clear, clc, close all

temp = load('icecore_temperature_data.txt');
accum = load('icecore_snowaccumulation_data.txt');
```

We then generate a simple line graph of the temperature data `temp` and snow accumulation data `accum`, as demonstrated in ►Chap. 4.

```
line1 = line(temp(:,1),temp(:,2));
```

This command opens a *Figure Window* named *Figure 1*, with a gray background, an x-axis ranging from 0 to 50 kyr, a y-axis ranging from -55 to -25 °C, and a blue line. Advanced editing of graphics requires an understanding of the MATLAB requirement for graphics to be organized as hierarchical suites of graphics objects. Put simply, the hierarchies of graphical objects include four main layers: *Root, Figure, Axes,* and *Chart and Primitive Objects*. The objects within each layer have a fixed set of properties, most of which can be modified to alter the appearance of a graph. In our example the line graph consists of a root object, a figure object, an axes object, and a chart line object. The values of each set of properties can be queried and most of these values can be modified. The values for the properties of the chart line object can be queried by typing

```
line1
```

which yields

```
line1 =
  Line with properties:
               Color: [0 0.4470 0.7410]
           LineStyle: '-'
           LineWidth: 0.5000
              Marker: 'none'
          MarkerSize: 6
     MarkerFaceColor: 'none'
               XData: [1x1632 double]
               YData: [1x1632 double]
               ZData: [1x0 double]
    Show all properties
```

Clicking on *Show all properties* while the window is still open yields a list of additional properties of the chart line object. Individual properties such as the property `LineWidth` can be accessed and changed using

```
line1.LineWidth
line1.LineWidth = 3;
```

This command changes the line width to a value of 3 pt (3 points).

The dot notation in `line1.LineWidth` to access and change graphics object properties was introduced with the R2014b release of MATLAB. With this release the numeric graphics object handles of type `double` were replaced by object handles of different types, depending on the class of the graphics object. Earlier versions of MATLAB used the `get` and `set` functions to access and modify graphics properties, instead of the dot notation. Fortunately, these functions still work with R2014b and later versions:

```
get(line1,'LineWidth')
set(line1,'LineWidth',3)
```

which yields

```
ans =
    0.5000

ans =
    3
```

displaying the default 0.5 pt and modified 3 pt line widths.

If we want to look at all four layers of the hierarchy of graphical objects, we start with `Root`, which includes information about the graphics environment and the current state of the graphics system, such as the display information and the pointer location. Herein, the function `root` lists the properties of the root object, which we store in `root1`. Some of these properties, such as `MonitorPosition`, which contains the size of the computer display in pixels, are read-only. After closing the previous figure window with

```
close all
```

we can use

```
root1 = groot
root1.MonitorPositions
root1.PointerLocation = [500 400]
```

to access the size of the display in pixels and move the pointer to the position [500 400] pixels on the display. Some properties, such as PointerLocation, are not available for all computer systems (e.g. for computers running *macOS*) and therefore does not change the pointer location on Macs. The Figure property includes the information about the appearance and behavior of a particular instance of a figure. As an example, we can use

```
figure1 = figure
figure1.Color
figure1.Color = [1 1 1]
```

to access and change the color of the background of the figure window from its default red, green, blue (RGB) values of [0.8 0.8 0.8], which is light gray, to values of [1 1 1], which is white. The fourth layer of graphics objects is axes, to control the appearance and behavior of the axes objects. We can use

```
axes1 = axes
axes1.XDir = 'reverse'
axes1.XLim = [0 20]
axes1.YLim = [-60 -20]
axes1.Box = 'on'
axes1.Title.String = 'GISP Ice Core Temperature Data'
axes1.XLabel.String = 'Age- Thousands of Years Before the Present'
axes1.YLabel.String = 'Temperature- Degrees Celsius'
```

to define the limits of the axes, with the *x*-axis being reversed, and to close the two open sides of the graph to form a box. In addition, we can define a title and apply labels to the axes. Alternatively, we can use the functions title, xlabel, and ylabel, as follows:

```
title1 = title('GISP Ice Core Temperature Data');
xlabel1 = xlabel('Age- Thousands of Years Before the Present');
ylabel1 = ylabel('Temperature- Degrees Celsius');
```

The function line controls the appearance and behavior of primitive line objects. We can use this function to access the default line color, which is [0 0.4470 0.7410], and change it to [0.1 0.3 0.8] using

```
line1 = line(temp(:,1),temp(:,2))
line1.Color = [0.1 0.3 0.8]
```

We can also use these commands to modify any graphic property in a single set of commands, as follows:

```
figure1 = figure('Color',[1 1 1]);
axes1 = axes('XLim',[0 20],...
    'XDir','reverse',...
    'YLim',[-60 -20],...
```

```
    'Box','on');
line1 = line(temp(:,1),temp(:,2),...
    'Color',[0.1 0.3 0.8]);
title1 = title('GISP Ice Core Temperature Data');
xlabel1 = xlabel('Age- Thousands of Years Before the Present');
ylabel1 = ylabel('Temperature- Degrees Celsius');
```

We now wish to plot both the temperature and the snow accumulation data on a single plot, each represented by different colored lines, using the function `yyaxis`. Together with `line`, this function creates plots with different y-axes on the left and the right of the graph, each with different tick labels and with the axis labels and colors corresponding to the colors of the lines. Again, we first clear the workspace and Command Window, close all Figure Windows, and reload the data using

```
clear, clc, close all
temp = load('icecore_temperature_data.txt');
accum = load('icecore_snowaccumulation_data.txt');
```

A simple chart with two y-axes can be created using

```
yyaxis left
line(temp(:,1),temp(:,2))
yyaxis right
line(accum(:,1),accum(:,2))
```

If the axes already exist, then the command `yyaxis left` activates the left y-axis, and `yyaxis right` activates the right y-axis. If the current chart does not have two y-axes then this command adds a second y-axis, identical to the one on the left. We can use `figure`, `axes` and `line` to change the properties of these objects, as in the previous example. The functions `title`, `xlabel`, and `ylabel` are used to define a title and apply labels to the axes of the chart:

```
figure1 = figure('Color',[1 1 1]);
axes1 = axes('XLim',[0 20],...
    'XDir','reverse',...
    'YLim',[-60 -20],...
    'Box','on');
yyaxis(axes1,'left')
axes1.YLim = [-60 -20]
axes1.YColor = [0.8 0.3 0.1]
line1 = line(temp(:,1),temp(:,2),...
    'Color',[0.8 0.3 0.1])
title1 = title('GISP Ice Core Temperature and Accumulation Data')
xlabel1 = xlabel('Age- Thousands of Years Before the Present')
ylabel1 = ylabel('Temperature- Degrees Celsius')
yyaxis(axes1,'right')
axes1.YLim = [0 0.4]
axes1.YColor = [0.1 0.3 0.8]
line2 = line(accum(:,1),accum(:,2),...
    'Color',[0.1 0.3 0.8])
ylabel2 = ylabel('Snow Accumulation- meters/year')
```

In this book we try to display all graphs in a similar layout. Our graphics template has the following settings: the length of the x-axis is 10 cm, the length of the y-axis is 8 cm, the line thickness of the axes is 0.6 pt, the line thickness of curves is 0.5 pt, the axis labels and title are in a sans serif font such as *Helvetica*, the font size of the axis labels is 8 pt, and the font size of the title is 10 pt (◻Fig. 5.1). The corresponding MATLAB script reads as follows:

```
figure1 = figure('Color',[1 1 1]);
axes1 = axes('Units','Centimeters',...
    'Position',[3 2 10 8],...
    'LineWidth',0.6,...
    'FontName','Helvetica',...
    'FontSize',8,...
    'XLim',[0 20],...
    'XDir','reverse',...
    'YLim',[-60 -20],...
    'Box','on');
yyaxis(axes1,'left')
axes1.YLim = [-60 -20]
axes1.YColor = [0.8 0.3 0.1]
line1 = line(temp(:,1),temp(:,2),...
    'Color',[0.8 0.3 0.1])
title1 = title('GISP Ice Core Temperature and Accumulation Data',...
    'FontName','Helvetica',...
    'FontSize',10);
xlabel1 = xlabel('Age- Thousands of Years Before the Present',...
    'FontName','Helvetica',...
    'FontSize',8);
ylabel1 = ylabel('Temperature- Degrees Celsius',...
    'FontName','Helvetica',...
```

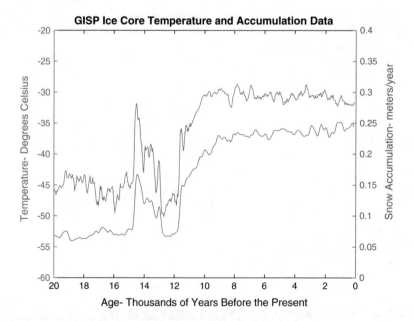

◻ **Fig. 5.1**   Line plots of temperature and snow accumulation variations from the GISP2 data (from Alley 2000) derived from the two files created in ► Chap. 3. In the original graph created with MATLAB, the temperature curve is orange (upper line) and the snow accumulation curve is blue (lower line)

```
    'FontSize',8);
yyaxis(axes1,'right')
axes1.YLim = [0 0.4]
axes1.YColor = [0.1 0.3 0.8]
line2 = line(accum(:,1),accum(:,2),...
    'Color',[0.1 0.3 0.8])
ylabel2 = ylabel('Snow Accumulation- meters/year',...
    'FontName','Helvetica',...
    'FontSize',8);
```

This script can be easily modified to suit to the reader's own requirements. Other graphics properties can also be included to change the final layout of the line graph. Having completed our line graph, we save the result in a file for further editing in a vector graphics program (see ▶Chap. 8).

```
print -depsc2 icecore_lineplot_vs1_matlab.eps
```

The figure window needs to be active when saving an image to a file using `print`. The preferred format to use for saving vector graphics such as the one in our example is the *Encapsulated Level 2 Color PostScript* (EPSC2) format. Note that the axes of the plot consist of several superimposed lines that cause color interferences. A possible way around this problem is described in ▶Chap. 8.

5.3 Bar Graphs: Plotting Histograms in Earth Sciences

Our second example demonstrates how to create bar graphs with MATLAB. Again, we use the temperature data from Alley (2000) contained in the file *icecore_temperature_data.txt*.

```
clear, clc, close all

temp = load('icecore_temperature_data.txt');
```

In statistical analyses, bar graphs (or plots), or histograms, are used to visualize frequency distributions of univariate data sets that have been organized into separate intervals, bins, or classes. We use a histogram to display the frequency distribution of the temperature values. The function `histogram` determines the bin edges using an automatic binning algorithm that returns uniform bins of a width that is chosen to cover the range of values in the data set. A number of binning algorithms are available with the function `histogram`; we can obtain a list of these algorithms with `help histogram`. Then, it counts the number of observations within each bin and displays the frequency distribution as a bar graph.

```
histogram(temp(:,2))
```

This command creates a histogram with bin edges at 2 °C intervals from −54 to −28 °C. The function `histogram` also returns a histogram object, to be used to inspect and adjust the properties of the histogram. Typing

```
histogram1 = histogram(temp(:,2))
```

5

yields

```
histogram1 =
  Histogram with properties:
            Data: [1632x1 double]
          Values: [38 86 138 82 150 82 87 62
                   33 48 84 623 119]
         NumBins: 13
        BinEdges: [-54 -52 -50 -48 -46 -44 -42
                   -40 -38 -36 -34 -32 -30 -28]
        BinWidth: 2
       BinLimits: [-54 -28]
   Normalization: 'count'
       FaceColor: 'auto'
       EdgeColor: [0 0 0]
  Show all properties
```

which displays the bin edges, the counts, and other properties of the histogram. Alternatively, we can calculate the bin centers v and the counts n from the histogram object histogram1 using

```
v = histogram1.BinWidth * 0.5 + histogram1.BinEdges(1:end-1);
n = histogram1.Values;
```

to display the bar plot of the frequency distribution in a similar layout to that described previously (◻Fig. 5.2) by typing

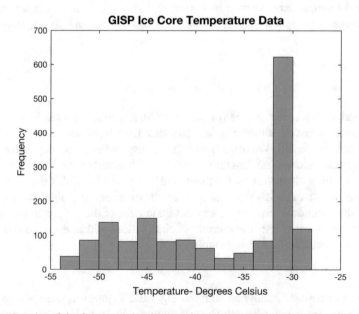

◻ **Fig. 5.2** Bar plot of the frequency distribution of temperature values from the GISP2 data (from Alley 2000) derived from the file created in ▶Chap. 3. In the original graph created with MATLAB, the histogram shows orange bars

```
figure1 = figure('Color',[1 1 1]);
axes1 = axes('Units','Centimeters',...
    'Position',[3 2 10 8],...
    'Box','on',...
    'XLim',[-55 -25],...
    'LineWidth',0.6,...
    'FontSize',8); hold on
bar1 = bar(axes1,v,n,...
    'BarWidth',1.0,...
    'LineWidth',0.5,...
    'FaceColor',[0.8 0.3 0.1],...
    'FaceAlpha',0.6);
title1 = title('GISP Ice Core Temperature Data',...
    'FontSize',10);
xlabel1 = xlabel('Temperature- Degrees Celsius',...
    'FontSize',8);
ylabel1 = ylabel('Frequency',...
    'FontSize',8);
```

Once again this script can be easily modified. Having completed our bar plot we save the result in a file for further editing in vector graphics software (see ▶ Chap. 8).

```
print -depsc2 icecore_bargraph_vs1_matlab.eps
```

As stated previously, the figure window needs to be active while saving the image to a file, and the preferred format to use for saving vector graphics such as the one in our example is the EPSC2 format.

5.4 Pie Charts: Illustrating Proportion in Earth Sciences

Our third example demonstrates how to create pie charts with MATLAB. Again, we use the temperature data from Alley (2000) contained in the file *icecore_temperature_data.txt*.

```
clear, clc, close all

temp = load('icecore_temperature_data.txt');
```

We then again calculate the frequency distribution of the temperature data:

```
histogram1 = histogram(temp(:,2));
n = histogram1.Values;
```

We can now create a pie chart of the frequency distribution. This chart does not provide any information on the actual values that contribute to each slice of the pie.

```
pie1 = pie(n);
```

The function `pie` basically scales the data to 100% and plots it as slices of a pie. The total number of temperature values in our example is $sum(n)=1632$. All values in n are therefore divided by 1632 and multiplied by 100 to obtain percentages. As you can see, the largest value is 38.1740, which is rounded to the nearest integer (38) for the plot label. The function `pie` returns a vector `pie1` containing patch and text handles.

```
percentages = 100*n/1632
```

The function `pie` also allows some slices to be offset from the center, using the vector `explode` to identify the slices to be offset by marking them with a `1`, while the slices that are not offset are marked with a `0`.

```
explode = zeros(size(n));
explode(1,5:7) = [1 1 1];
pie1= pie(n,explode);
```

5

We can also plot the frequency distribution in a 3D pie chart. This will be the first 3D plot that we create in this book. We will learn more about 3D charts in the next chapter.

```
pie1 = pie3(n);
```

We again display the graphs in our standard layout. The corresponding MATLAB script reads as follows:

```
figure1 = figure('Color',[1 1 1]);
axes1 = axes('Units','Centimeters',...
    'Position',[3 2 10 8],...
    'LineWidth',0.6,...
    'FontSize',8);
pie1 = pie3(n);
```

As before, this script can be easily modified. Having completed our bar plot, we save the result in a file for further editing in vector graphics software.

```
print -depsc2 icecore_piechart_vs1_matlab.eps
```

Again, the figure window needs to be active while saving the image to a file and the preferred format to use for saving vector graphics such as the one in our example is the EPSC2 format.

A useful feature of MATLAB graphics is that of transparency. We use the function `alpha` to make the pie chart transparent, using a value of `0.3` (◼Fig. 5.3), where `1` is opaque and `0` is clear.

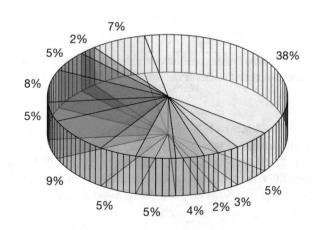

◼ **Fig. 5.3** Pie chart of the temperature data from Alley (2000) contained in the file *icecore_temperature_data.txt*. We use the function `alpha` to make the pie chart transparent with a value of `0.3`, where `1` is opaque and `0` is clear. The original graph created with MATLAB shows colors

```
figure1 = figure('Color',[1 1 1]);
axes1 = axes('Units','Centimeters',...
    'Position',[3 2 10 8],...
    'LineWidth',0.6,...
    'FontSize',8);
pie1 = pie3(n);
alpha(0.3)
```

Unfortunately, graphs using transparency are not exported as vector graphics. Trying to export the pie chart as an EPS file creates a raster or pixel version of the graph.

```
print -depsc2 icecore_piechart_vs2_matlab.eps
```

We therefore export the transparent pie chart as a JPEG file

```
print -djpeg70 -r600 icecore_piechart_vs3_matlab.jpg
```

with a quality level of 70% and a resolution of 600 dpi. Alternatively, we can export the non-transparent version of the pie chart and introduce transparency using vector graphics software (▶Chap. 8).

5.5 Rose Diagrams: Plotting Directional Data

The classic way to display directional data is the rose diagram. A rose diagram is a histogram for angle measurements. In contrast to a bar histogram, in which the height of the bars is proportional to the frequency, a rose diagram comprises sectors of a circle in which the radius of each sector is proportional to the frequency. We use synthetic data to illustrate two types of rose diagram for displaying directional data. We first load a set of directional data from the file *directional_data.txt*.

```
clear, clc, close all

data_degrees_1 = load('directional_data.txt');
```

This data set contains forty measurements of angles, in degrees. We use the function `rose(az,nb)` to display the data. This function plots an angle histogram for the angles `az` in radians, where `nb` is the number of classes. However, since the original data are in degrees, we need to convert all measurements to radians before we plot the data.

```
data_radians_1 = pi*data_degrees_1/180;
rose(data_radians_1,12)
```

The function `rose` counts in a counterclockwise direction, with zero degrees lying along the *x*-axis of the coordinate graph. In geosciences, however, the opposite applies and degrees are counted in a clockwise direction from due north, which is zero degrees. The command `view` rotates the plot by +90° (the azimuth) and mirrors the plot by −90° (the elevation).

```
rose(data_radians_1,12)
view(90,-90)
```

■ **Fig. 5.4** Rose diagram of directional data from the file *directional_data.txt*. In the original graph created with MATLAB, the arc segments have an orange line color

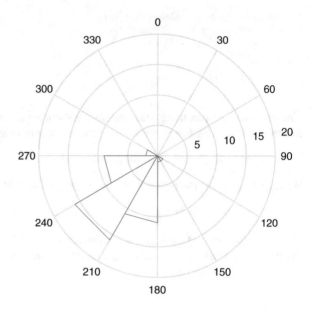

We again display the graph in our standard layout (■Fig. 5.4). The corresponding MATLAB script reads as follows:

```
figure1 = figure('Color',[1 1 1]);
axes1 = axes('Units','Centimeters',...
    'Position',[3 2 10 8],...
    'LineWidth',0.6,...
    'FontSize',8);
rose1 = rose(data_radians,12);
rose1.Color = [0.8 0.3 0.1]
view(90,-90)
```

As before, this script can be easily modified. Having completed our rose diagram, we save the result in a file for further editing in vector graphics software (see ▶Chap. 8).

```
print -depsc2 icecore_rosediagram_vs1_matlab.eps
```

Once again, the figure window needs to be active while saving the image to a file and the preferred format to use for saving vector graphics such as the one in our example is the EPSC2 format.

5.6 Multiplots: Plotting Scaled Multiple Area Graphs

This specialized example creates several plots in a single figure window. It plots patch, line, or bar plots of several pollen or microfossil records, all on axes with the same scales. We first need to create some synthetic data. The first column contains the age vector, while all other data are uniformly-distributed pseudo random numbers. The total of all values within each column is 100. We can change the number of data points m.

```
clear, clc, close all

rng(10)
m = 40;
data(:,1)  = sortrows(m*rand(m,1));
data(:,2:15) = rand(m,14);
data(:,2)  = 5*data(:,2);
data(:,3)  = 5*data(:,3);
data(:,4)  = 10*data(:,4);
data(:,5)  = 5*data(:,5);
data(:,6)  = 5*data(:,6);
data(:,7)  = 3*data(:,7);
data(:,8)  = 5*data(:,8);
data(:,9)  = 5*data(:,9);
data(:,10) = 15*data(:,10);
data(:,11) = 12*data(:,11);
data(:,12) = 5*data(:,12);
data(:,13) = 3*data(:,13);
data(:,14) = 2*data(:,14);
data(:,15) = 2*data(:,15);
data(1,1)  = 0;
```

We also create a list of fossil names by typing

```
names = ["Lorem ipsum"
         "Dolor sit"
         "Amet consectetuer"
         "Adipiscing elit"
         "Aenean commodo"
         "Ligula eget"
         "Dolor aenean"
         "Massa cum"
         "Sociis natoque"
         "Penatibus et"
         "Magnis dis"
         "Parturient montes"
         "Nascetur ridiculus"
         "Mus donec"
         "Quam felis"
         ];
```

The synthetic data is then saved to the binary file *multipledata_data.mat*.

```
save multipledata_data.mat data
```

We now clear the workspace and import the data from the binary file *multipledata_data.mat*.

```
clear
```

```
load multipledata_data.mat
```

Alternatively, we can import any other data set with an arbitrary number of data points and variables, provided the total comes to 100%. We add an age of zero at both ends of the time series by typing

5

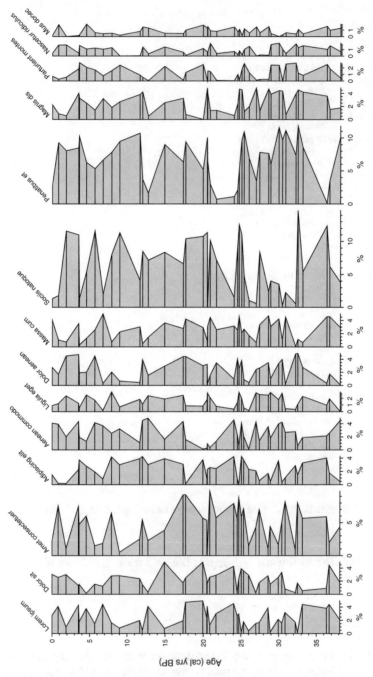

■ **Fig. 5.5** Multiple patch plots displaying pollen and microfossil records, all plotted at the same scale. In the original graph created with MATLAB, the patches have an orange color that can be easily modified for other applications

```
[m,n] = size(data);
data(2:m+1,:) = data;
data(1,:) = zeros(1,n);
data(1,1) = data(2,1);
data(m+2,:) = zeros(1,n);
data(m+2,1) = data(m+1,1);
```

The `multiplot` function first creates a figure window with the same size as the screen. It then plots all records at the same scale (◘Fig. 5.5).

```
root1 = groot;
scrsz = root1.ScreenSize;
scrsz = 0.6*scrsz;
sf = 0.75;
colorcode = [0.8 0.8 0.4];
figure1 = figure(...
    'Position',[1 scrsz(4) scrsz(3) scrsz(4)],...
    'Color',[1 1 1]);
for i = 1:n-1
    if i == 1
    xcoord = max(data(:,2))/100;
    axes1 = axes(...
        'Parent',figure1,...
        'YMinorTick','on',...
        'XMinorTick','on',...
        'TickDir','out',...
        'Position',[0.1 0.1 sf*xcoord 0.6],...
        'FontSize',12,...
        'XLim',[0 max(data(:,2))],...
        'YLim',[min(data(:,1)) max(data(:,1))],...
        'YDir','reverse',...
        'Box','off');
    patch1 = patch(data(:,2),data(:,1),colorcode); hold on
    %line1 = line(data(:,2),data(:,1),...
    %       'Color',[0 0 0]);
    barh1 = barh(data(2:end-1,1),data(2:end-1,2),...
        'BarWidth',0,...
        'EdgeColor',[0 0 0],...
        'FaceColor',colorcode);
    title1 = title(names(i),...
        'Rotation',45,...
        'FontWeight','normal',...
        'Position',[0 -1],...
        'HorizontalAlignment','left',...
        'VerticalAlignment','top',...
        'FontSize',12);
    xlabel1 = ylabel('Age (cal yrs BP)');
    ylabel1 = xlabel('%');
    xcoord = xcoord + 0.01;
    else
    axes2 = axes(...
        'Parent',figure1,...
        'YTick',zeros(1,0),...
        'FontSize',12,...
        'YMinorTick','on',...
        'XMinorTick','on',...
        'TickDir','out',...
        'Position',[sf*xcoord+0.1 0.1 ...
            sf*max(data(:,i+1))/100 0.6],...
        'XLim',[0 max(data(:,i+1))],...
```

```
        'YLim',[min(data(:,1)) max(data(:,1))],...
        'YDir','reverse',...
        'Box','off');
    patch2 = patch(data(:,i+1),data(:,1),colorcode); hold on
    %line2 = line(data(2:end-1,i+1),data(2:end-1,1),...
    %     'Color',[0 0 0]);
    barh2 = barh(data(2:end-1,1),data(2:end-1,i+1),...
        'BarWidth',0,...
        'EdgeColor',[0 0 0],...
        'FaceColor',colorcode);
    xlabel2 = xlabel('%');
    title(names(i),...
        'Rotation',45,...
        'FontWeight','normal',...
        'Position',[0 -1],...
        'HorizontalAlignment','left',...
        'VerticalAlignment','top',...
        'FontSize',12);
    xcoord = xcoord + max(data(:,i+1))/100 + 0.01;
    end
end
```

We can change the multiplier for `scrsz` between 0 and 1 to match the figure size with the screen size. We can also change `colorcode` to modify the color of the patches. The variable `sf` scales the graph with respect to the figure window. Instead of a `patch` plot, the graphics function `line` for a line plot is also available. To use either of these we need to uncomment the `patch` or the `line` command by removing the % signs at the beginning of the corresponding lines.

As in all previous examples, this script can be easily modified. Having completed our plot, we save the result in a file for further editing in vector graphics software (see ►Chap. 8).

```
print -depsc2 multipledata_multiplot_vs1_matlab.eps
```

As previously stated, the figure window needs to be active when saving the image to a file and the preferred format to use for saving vector graphics such as the one in our example is the EPSC2 format.

5.7 Stratplots: Plotting Stratigraphic Columns

The second specialized example introduces a MATLAB script to draw a stratigraphic column with either angular or rounded corners, labels, and a number of data curves. We first load the data from a binary file. The file contains the data matrix and a variable containing labels such as rock types.

```
clear, clc, close all

load stratigraphiccolumn_data.mat
```

We need to flip the data array `data` and the label array `litho` in an up/down direction if the first line in the data array represents the lowermost sediment layer, as this needs to be displayed at the bottom of the graph.

```
data = flipud(data);
litho = flipud(litho);
```

Next, we add a zero to either end in order to close the patches in the plot.

```
[n,m] = size(data);
datanew(1,:) = zeros(1,m);
for i = 1:n
    datanew(i+1,:) = data(i,:);
end
data = datanew;
data(n+2,:) = zeros(1,m);
```

We define specific variables, such as t for the layer thickness, w for the degree of weathering, l for the type of lithology, and ct for the cumulative thickness, to plot the data.

```
t = data(:,1);
w = data(:,2);
wc = data(:,2)./max(data(:,2));
l = data(:,3);
lc = data(:,3)./max(data(:,3));
ct = cumsum(t);
```

Finally, we display the stratigraphic column as a sequence of rectangular colored layers in which the color represents the sediment type. The script first creates a white figure window, then creates a stratigraphic column with angular corners using fill, labels the layers, and creates two curves using the function patch.

```
figure1 = figure('Color',[1 1 1]);

% PART 1 Stratigraphic column

axes1 = axes('position',[0.15 0.1 0.1 0.8],...
    'YLim',[0 max(ct)],...
    'Color','None',...
    'TickDir','out',...
    'XTickLabel',[],...
    'XTick',[]);
hold on
for i=2:length(ct)
    layercolor = [0.8 lc(i) 0.1];
    X=[0 0 w(i) w(i)];
    Y=[ct(i-1) ct(i) ct(i) ct(i-1)];
    fill1 = fill((X).^.5,Y,layercolor);
end
hold off

% PART 2: Labels

axes1 = axes('position',[0.25 0.1 0.1 0.8],...
    'YLim',[0 max(ct)],...
    'Color','None',...
    'Visible','off');
for i=1:length(ct)-2
    text(0.35,ct(i+1)-t(i+1)./2,litho(i,:),...
        'Fontsize',6)
end
```

```
% PART 3: Patch 1

axes1 = axes('position',[0.4 0.1 0.1 0.8],...
    'YLim',[0 max(ct)],...
    'Color','None',...
    'TickDir','out',...
    'XTickLabel',[],...
    'XTick',[]);
patch1 = patch(data(:,4),ct,[0.8 0.3 0.1]);

% PART 4: Patch 2

axes1 = axes('position',[0.55 0.1 0.1 0.8],...
    'YLim',[0 max(ct)],...
    'Color','None',...
    'TickDir','out',...
    'XTickLabel',[],'XTick',[]);
patch1 = patch(data(:,5),ct,[0.8 0.3 0.1]);
```

Having completed our plot, we save the result in a file for further editing in vector graphics software (see ▶Chap. 8).

```
print -depsc2 stratigraphiccolumn_stratplot_vs1_matlab.eps
```

Once again, the figure window needs to be active while saving the image to a file and the preferred format to use for saving the vector graphics such as the one in our example is again the EPSC2 format.

A modification of the previous graph is a stratigraphic column with rounded corners to the layers. The script first creates a white figure window, then creates a stratigraphic column with rounded corners using `fill`, labels the layers, and creates two curves using the function `patch`. The script is slightly more complicated than that for the angular version since the rounded corners differ depending on the width shown for the overlying and underlying layers. The rounding is either convex or concave depending on the neighboring layer, or not rounded if the neighboring layers have the same width. The script for labeling the layers and for creating the two curves using `patch` is very similar to those described in the previous sections (◻Fig. 5.6).

```
figure1 = figure('Color',[1 1 1]);

% PART 1 Stratigraphic column

axes1 = axes('position',[0.15 0.1 0.1 0.8],...
    'YLim',[0 max(ct)],...
    'Color','None',...
    'TickDir','out',...
    'XTickLabel',[],...
    'XTick',[]);
hold on
for i=2:length(ct)-1
  layercolor = [0.8 lc(i) 0.1];
if  w(i)==2&w(i-1)==1&w(i+1)==1
  X=[0 0 w(i)-0.2 w(i)  w(i)  w(i)-0.2];
  Y=[ct(i-1) ct(i) ct(i) ct(i)-0.2*mean(t) ...
      ct(i-1)+0.2*mean(t) ct(i-1)];
  fill1 = fill((X).^.5,Y,layercolor);
elseif w(i)==1 & w(i-1)==2 & w(i+1)==2
```

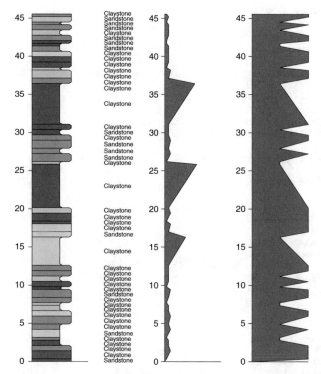

◘ Fig. 5.6 Stratigraphic plot using a MATLAB script to draw a stratigraphic column with rounded corners, labels and several data curves. In the original graph created with MATLAB, the different stratigraphic units are in different colors

```
X=[0 0 w(i)+0.2 w(i) w(i) w(i)+0.2];
Y=[ct(i-1) ct(i) ct(i) ct(i)-0.2*mean(t) ...
    ct(i-1)+0.2*mean(t) ct(i-1)];
fill1 = fill((X).^.5,Y,layercolor);
elseif w(i)==1&(w(i-1)==1&w(i+1)==1)|(w(i-1)==0&w(i+1)==1)
   X=[0 0 w(i) w(i)];
   Y=[ct(i-1) ct(i) ct(i) ct(i-1)];
   fill1 = fill((X).^.5,Y,layercolor);
elseif w(i)==1&w(i-1)==1&w(i+1)==0
   X=[0 0 w(i) w(i)];
   Y=[ct(i-1) ct(i) ct(i) ct(i-1)];
   fill1 = fill((X).^.5,Y,layercolor);
elseif w(i)==2&(w(i-1)==2&w(i+1)==2)|(w(i-1)==0&w(i+1)==2)
   X=[0 0 w(i) w(i)];
   Y=[ct(i-1) ct(i) ct(i) ct(i-1)];
   fill1 = fill((X).^.5,Y,layercolor);
elseif w(i)==2&w(i-1)==2&w(i+1)==0
   X=[0 0 w(i) w(i)];
   Y=[ct(i-1) ct(i) ct(i) ct(i-1)];
   fill1 = fill((X).^.5,Y,layercolor);
elseif w(i)==2&(w(i-1)==2&w(i+1)==1)|(w(i-1)==0&w(i+1)==1)
   X=[0 0 w(i)-0.2 w(i) w(i)];
   Y=[ct(i-1) ct(i) ct(i) ct(i)-0.2*mean(t) ct(i-1)];
   fill1 = fill((X).^.5,Y,layercolor);
elseif w(i)==2&(w(i-1)==1&w(i+1)==2)|(w(i-1)==1&w(i+1)==0)
```

```
    X=[0  0  w(i)  w(i)  w(i)-0.2];
    Y=[ct(i-1)  ct(i)  ct(i)  ct(i-1)+0.2*mean(t)  ct(i-1)];
    fill1 = fill((X).^.5,Y,layercolor);
elseif w(i)==1&w(i-1)==1&w(i+1)==2
    X=[0  0  w(i)+0.2  w(i)  w(i)];
    Y=[ct(i-1)  ct(i)  ct(i)  ct(i)-0.2*mean(t)  ct(i-1)];
    fill1 = fill((X).^.5,Y,layercolor);
elseif w(i)==1&w(i-1)==2&w(i+1)==1
    X=[0  0  w(i)  w(i)  w(i)+0.2];
    Y=[ct(i-1)  ct(i)  ct(i)  ct(i-1)+0.2*mean(t)  ct(i-1)];
    fill1 = fill((X).^.5,Y,layercolor);
end
end

% PART 2: Labels

axes1 = axes('position',[0.25 0.1 0.1 0.8],...
    'YLim',[0 max(ct)],...
    'Color','None',...
    'Visible','off');
for i=1:length(ct)-2
    text1 = text(0.35,ct(i+1)-t(i+1)./2,litho(i,:),...
        'Fontsize',6);
end

% PART 3: Patch 1

axes1 = axes('position',[0.4 0.1 0.1 0.8],...
    'YLim',[0 max(ct)],...
    'Color','None',...
    'TickDir','out',...
    'XTickLabel',[],...
    'XTick',[]);
patch1 = patch(data(:,4),ct,[0.8 0.3 0.1]);

% PART 4: Patch 2

axes1 = axes('position',[0.55 0.1 0.1 0.8],...
    'YLim',[0 max(ct)],...
    'Color','None',...
    'TickDir','out',...
    'XTickLabel',[],'XTick',[]);
patch1 = patch(data(:,5),ct,[0.8 0.3 0.1]);
```

Having completed our plot, we save the result in a file for further editing in vector graphics software (see ▶Chap. 8).

```
print -depsc2 stratigraphiccolumn_stratplot_vs2_matlab.eps
```

As always, the figure window needs to be active while printing the image to a file and the preferred format to use for saving vector graphics such as the one in our example is again the EPSC2 format.

Recommended Reading

Alley RB (2000) The Younger Dryas cold interval as viewed from central Greenland. Quaternary Science Reviews 19 (1–5):213–226

Visualizing 3D Data in Earth Sciences

Electronic Supplementary Material The online version of this chapter (https://doi.org/10.1007/978-3-662-56203-1_6) contains supplementary material, which is available to authorized users.

© Springer-Verlag GmbH Germany, part of Springer Nature 2018
M. H. Trauth and E. Sillmann, *Collecting, Processing and Presenting Geoscientific Information*, Springer Textbooks in Earth Sciences, Geography and Environment, https://doi.org/10.1007/978-3-662-56203-1_6

6.1 Introduction

Most data in earth sciences are spatially distributed, either as *vector data*, (points, lines, polygons) or as *raster data* (gridded topography). Vector data are generated by digitizing map objects such as drainage networks or outlines of lithologic units. Raster data can be obtained directly from a satellite sensor output, but gridded data can also, in most cases, be interpolated from irregularly-distributed field samples (gridding).

This chapter demonstrates advanced three-dimensional visualization techniques through graphical displays of the types of data typically encountered in earth sciences, using MATLAB. The following section introduces the use of vector data by using coastline data as an example (▶Sect. 6.2). The acquisition and handling of raster data are then illustrated using digital topographic data (▶Sects. 6.3–6.5). The availability and use of digital elevation data has increased considerably since the early 1990s. With a resolution of 5 arc minutes (about 9 km), ETOPO5 was one of the first data sets for topography and bathymetry. In October 2001 it was replaced by Etopo1, which has a resolution of 2 arc minutes (about 4 km), and in March 2009 the ETOPO1 became available, which has a resolution of 1 arc minutes (about 2 km). There is also a data set for topography called GTOPO30, completed in 1996, that has a horizontal grid spacing of 30 arc seconds (about 1 km). More recently, the 30 m resolution data from the Shuttle Radar Topography Mission (SRTM) have replaced the older data sets in most scientific studies. In contrast to airborne data, most land-based earth science data are collected on irregular sampling patterns. Access to rocks is often restricted to natural outcrops such as shoreline cliffs and the walls of a gorge, or anthropogenic outcrops such as road cuttings and quarries. ▶Section 6.6 is on interpolating such unevenly-spaced data and displaying the results in two- and three-dimensional graphs.

This chapter requires the Mapping Toolbox (MathWorks 2016), although most graphics routines used in our examples can be easily replaced by standard MATLAB functions. An alternative and useful mapping toolbox by Rich Pawlowicz (Earth and Ocean Sciences, at the University of British Columbia) is available from

```
http://www.eos.ubc.ca/~rich/
```

The handling and processing of large spatial data sets also requires a computing system with at least 4 GB physical memory.

6.2 The Global Geography Database GSHHG

The *Global Self-consistent, Hierarchical, High-resolution Geography* (GSHHG) database is an amalgamation of two public domain databases by Paul Wessel (SOEST, University of Hawaii, Honolulu, HI) and Walter Smith (NOAA Laboratory for Satellite Altimetry, Silver Spring, MD) (Wessel and Smith 1996). The GSHHG database consists of the older GSHHS shoreline database (Soluri and Woodson 1990; Wessel and Smith 1996), which is a shoreline database, with the poor quality Antarctica data replaced by the more accurate data from Bohlander and Scambos (2007), and with rivers and borders taken from the CIA World Data Bank II (WDBII) (Gorny 1977). The GSHHG data can be downloaded from the SOEST server

```
http://www.soest.hawaii.edu/pwessel/gshhg/index.html
```

in three different formats: Generic Mapping Tools (GMT) files, ESRI shapefiles, or native binary files, and in a wide range of spatial resolutions. This example uses version 2.3.6 of the native binary file *gshhg-bin-2.3.6.zip*. We use the MATLAB unzip function to decompress this file

```
clear, clc, close all

unzip('gshhg-bin-2.3.6.zip')
```

which creates files containing the shorelines (*gshhs*), borders (*wdb_borders*) and rivers (*wdb_rivers*), each at different resolutions ranging from *f* (full resolution, 95.8 MB large) to *c* (crude resolution, 183 kb large), as explained on Paul Wessel's SOEST webpage. As an example we import the high-resolution shoreline data set from file *gshhs_h.b* using

```
S = gshhs('gshhs_h.b');
```

We obtain a structure array S which includes the longitude/latitude coordinates of the shoreline segments together with various other types of information. The elements of S contain the shorelines of the continents and islands, for example S(2) contains the longitude/latitude coordinates of Africa. We can extract the shoreline coordinates by typing

```
data(:,1) = S(2).Lon;
data(:,2) = S(2).Lat;
```

We save the longitude/latitude coordinates in an ASCII file using

```
save coastline_data.txt data -ascii
```

and load it after clearing the workspace using

```
data = load('coastline_data.txt');
```

There are different ways of displaying the data. As an example, we can use the line command to create a line plot, after having defined the figure and axes properties, as follows:

```
figure1 = figure('Color',[1 1 1]);
axes1 = axes('Box','on',...
    'DataAspectRatio',[1 1 1]);
line1 = line(data(:,1),data(:,2),...
    'Color',[0 0 0]);
xlabel1 = xlabel('Longitude');
ylabel1 = ylabel('Latitude');
```

More advanced plotting functions are contained in the Mapping Toolbox, which allows alternative versions of this plot to be generated:

```
axesm('MapProjection','lambert', ...
       'MapLatLimit',[0 15], ...
       'MapLonLimit',[35 55], ...
       'Frame','on', ...
       'MeridianLabel','on', ...
       'ParallelLabel','on');
plotm(data(:,2),data(:,1),'k');
```

Note that the input for `plotm` is given in the order *longitude*, followed by the *latitude*, i.e., the second column of the data matrix is entered first. In contrast, the function `plot` requires an *xy* input, i.e., the first column is entered first. The function `axesm` defines the map's axes and sets various map properties such as the map projection, the map limits and the axis labels. Alternatively, we can use the functions `worldmap` and `geoshow` from the Mapping Toolbox, as follows:

```
figure1 = figure('Color',[1 1 1]);
worldmap1 = worldmap('africa');
geoshow1 = geoshow(data(:,2),data(:,1),...
      'Color',[0 0 0]);
```

We then display the graphs in a similar layout to that used in the previous chapter. The corresponding MATLAB script reads as follows:

```
figure1 = figure('Color',[1 1 1]);
axes1 = axes('Units','Centimeters',...
       'Position',[3 2 10 8],...
       'Box','on',...
       'LineWidth',0.6,...
       'FontSize',8);
line1 = line(data(:,1),data(:,2),...
       'Color',[0.1 0.3 0.8],...
       'LineWidth',0.5);
title1 = title('GSHHS Shoreline Data Set',...
       'FontSize',10);
xlabel1 = xlabel('Longitude',...
       'FontSize',8);
ylabel1 = ylabel('Latitude',...
       'FontSize',8);
```

The length of the *x*-axis is 10 cm, the length of the *y*-axis is 8 cm, the line thickness of the axes is 0.6 pt (0.6 points), the line thickness of the shorelines is 0.5 pt, the axis labels and title are in a sans serif font such as *Helvetica*, the font size of the axis labels is 8 pt and that of the title is 10 pt (◘Fig. 6.1).

The shoreline data set is a polygon with 28,033 vertices. This large number causes problems when processing the polygon using vector graphics software. A maximum of 8,000 vertices, but preferably only 4,000, can be processed. To demonstrate the effect of large numbers of vertices, we create a vector graph of the full data set of 28,033 vertices, and then we decimate the data set by a factor of 2 and a factor of 4, for comparison. Decimating the data set by a factor of two can be done by using

```
data = data(1:2:end,:);
```

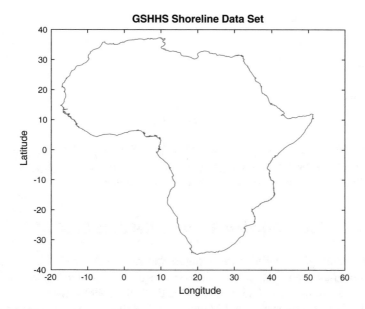

Fig. 6.1 Display of the GSHHS shoreline data set of Africa. This simple map is made using the function `line`

Pairs of `NaNs` may be present within the data set, representing pen up/pen down breaks of the polygon. Simply decimating the data set by `data = data(1:2:end,:)` would eliminate some of the `NaNs` and might therefore link polygons representing islands and lakes. A simple script overcomes this problem by creating a second data set `newdata` in which the data is shifted by one data point with respect to the original data set. We then have two data sets with neighboring `NaNs`. Replacing the real data point in `data` by a `NaN` if there is one in `newdata` produces neighboring pairs of `NaNs` in the original data set `data`. We can then decimate the data set by using `data(1:2:end,:)`. To decimate the data set by a factor of 4, we run the script a second time.

```
newdata = circshift(data,-1);
data(isnan(newdata) ==1) = NaN;
data = data(1:2:end,:);
```

The shorelines can again be displayed by using

```
figure1 = figure('Color',[1 1 1]);
axes1 = axes('Units','Centimeters',...
    'Position',[3 2 10 8],...
    'Box','on',...
    'LineWidth',0.6,...
    'FontSize',8);
line1 = line(data(:,1),data(:,2),...
    'Color',[0.1 0.3 0.8],...
    'LineWidth',0.5);
title1 = title('GSHHS Shoreline Data Set',...
```

```
        'FontSize',10);
xlabel1 = xlabel('Longitude',...
        'FontSize',8);
ylabel1 = ylabel('Latitude',...
        'FontSize',8);
```

The command

```
print -depsc2 coastline_linegraph_vs1_matlab.eps
```

creates an *Encapsulated Level 2 Color PostScript* (EPSC2) file, which is the preferred format to use for exporting vector graphics that contain line-drawings only, with no special effects such as interpolated shading or transparency.

6.3 The 1-Minute Gridded Global Relief Data ETOPO1

The *1 arc-minute global relief model of the Earth's surface* (ETOPO1) is a global database of topographic and bathymetric data on a regular 1 arc-minute grid (about 2 km) (Amante and Eakins 2009). Older ETOPO2 and ETOPO5 global relief grids have been superseded but are still available. ETOPO1 is a compilation of data from a variety of sources. It can be downloaded from the NOAA National Centers for Environmental Information (NCEI) webpage

```
https://www.ngdc.noaa.gov/mgg/global/global.html
```

We can download either the whole-world grids, or custom grids for ice surface or bedrock. We create and extract a custom grid online from ETOPO1 using NCEI's WCS Client:

```
http://maps.ngdc.noaa.gov/viewers/wcs-client/
```

We first choose the *ETOPO1 (bedrock)* layer. We then select *Select with Coordinates* by clicking the *xy* icon in the upper-left corner of the map and specify an area of interest; for example, a latitude from *North: 20* (20°N, or 20° northern latitude) and *South: −20* (20°S) and a longitude between *West: 30* (30°E) and *East: 60* (60°E) covers the East African coast. Negative latitude and longitude values are assigned for the southern hemisphere and for the western hemisphere. Clicking *OK* marks the area of interest as a transparent yellow rectangle on the map. Next, we choose ArcGIS ASCII Grid as the Output Format and download the file *etopo1_bedrock.asc*, which has the following content:

```
ncols        1801
nrows        2401
xllcorner    29.991666666667
yllcorner    -20.008333333333
cellsize     0.016666666667
294 299 293 288 285 282 ...
237 241 245 266 264 274 ...
259 263 267 262 263 266 ...
310 306 300 294 296 291 ...
348 346 352 356 353 353 ...
381 383 381 381 382 381 ...
(cont'd)
```

The headers document the size of the data matrix (e.g., 1801 columns and 2401 rows in our example), the coordinates of the lower-left corner (e.g., approximately $x = 20$ and $y = -20$), and the cell size (e.g., ~0.0167 = 1/60 degree latitude and longitude, or 1 arc minute). Older versions of the ETOPO1 data set, as well as other similar data sets, may also contain the -32768 flag for data voids. We comment the header by typing % at the beginning of the first five lines

```
%ncols         1801
%nrows         2401
%xllcorner     29.991666666667
%yllcorner     -20.008333333333
%cellsize       0.016666666667
294 299 293 288 285 282 ...
237 241 245 266 264 274 ...
259 263 267 262 263 266 ...
310 306 300 294 296 291 ...
348 346 352 356 353 353 ...
381 383 381 381 382 381 ...
(cont'd)
```

save the file as *etopo1_data.txt*, and then load the data into the workspace.

```
clear, clc, close all

ETOPO1 = load('etopo1_data.txt');
```

We flip the matrix up and down to change the indexing of the data array according to MATLAB conventions. The -32768 flag for data voids, if any, must then be replaced by the MATLAB representation for Not-a-Number NaN.

```
ETOPO1 = flipud(ETOPO1);

ETOPO1(find(ETOPO1 == -32768)) = NaN;
```

Finally, we check that the data are now correctly stored in the workspace by printing the minimum and maximum elevations for the area.

```
max(ETOPO1(:))
min(ETOPO1(:))
```

In this example the maximum elevation for the area is 5,677 m and the minimum elevation is $-5,859$ m. The reference level is the sea level at 0 m. We now define a coordinate system using the information that the lower-left corner is at latitude 20° south and longitude 30° east. The resolution is 1 arc minutes, corresponding to 1/60 degree.

```
[LON,LAT] = meshgrid(30:1/60:60,-20:1/60:20);
```

We now generate a colored surface from the elevation data using the function surf.

```
figure1 = figure('Color',[1 1 1]);
axes1 = axes('Box','on',...
    'View',[0 90]); hold on
surf1 = surf(LON,LAT,ETOPO1,...
    'FaceColor','interp',...
    'EdgeColor','none');
```

```
colorbar1 = colorbar;
```

This script opens a new figure window and generates a colored surface. The surface is highlighted by a set of color shades in an overhead view. We again display the graphs in a similar layout to that used in the previous chapter. The corresponding MATLAB script reads as follows:

```
figure1 = figure('Color',[1 1 1]);
axes1 = axes('Units','Centimeters',...
    'Position',[3 2 10 8],...
    'Box','on',...
    'View',[0 90],...
    'LineWidth',0.6,...
    'FontSize',8); hold on
surf1 = surf(LON,LAT,ETOPO1,...
    'FaceColor','interp',...
    'EdgeColor','none');
title1 = title('ETOPO1 Data Set',...
    'FontSize',10);
xlabel1 = xlabel('Longitude',...
    'FontSize',8);
ylabel1 = ylabel('Latitude',...
    'FontSize',8);
```

6

Details of the lines and fonts are as before. Furthermore, we set the axes property `view` to `[0,90]`. The command

```
print -depsc2 etopo1_pseudocolorplot_vs1_matlab.eps
```

exports the graph as an EPSC2 file. The result is obviously not a vector graph and we therefore export the graph as a JPEG file

```
print -djpeg70 -r600 etopo1_pseudocolorplot_vs2_matlab.jpg
```

compressed to 70%, with a resolution of 600 dpi.

We next create a filled contour plot of the data. The contour intervals are

```
v = [-6000 -4000 -2000 -1000 -500 0 500 1000 2000 4000 6000];
```

Again, the length of the x-axis is 10 cm, the length of the y-axis is 8 cm, the line thickness of the axes is 0.6 pt (0.6 points), the line thickness of curves is 0.5 pt, the font of the axis labels and title is a sans serif font such as Helvetica, the font size of the axis labels is 8 pt, and that of the title is 10 pt (◘Fig. 6.2).

```
figure1 = figure('Color',[1 1 1]);
axes1 = axes('Units','Centimeters',...
    'Position',[3 2 10 8],...
    'Box','on',...
    'View',[0 90],...
    'LineWidth',0.6,...
    'FontSize',8); hold on
contourf1 = contourf(LON,LAT,ETOPO1,v);
title1 = title('ETOPO1 Data Set',...
    'FontSize',10);
xlabel1 = xlabel('Longitude',...
    'FontSize',8);
ylabel1 = ylabel('Latitude',...
```

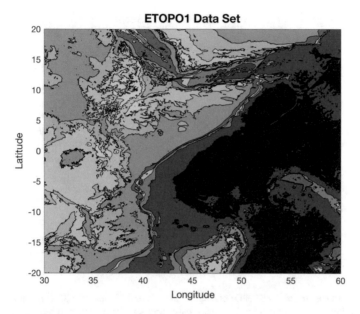

Fig. 6.2 Filled contour plot of the ETOPO1 elevation data. The function `demcmap` contained in the Mapping Toolbox creates and assigns a colormap appropriate for elevation data. According to the manual of the toolbox, the number of land and sea colors in the colormap is in proportion to the maximum elevations and depths in the matrix map

```
        'FontSize',8);
colormap1 = colormap(demcmap(ETOPO1));
```

The function `demcmap` contained in the Mapping Toolbox creates and assigns a colormap appropriate for elevation data. According to the toolbox manual the number of land and sea colors in the colormap is in proportion to the maximum elevations and depths in the matrix map. We can export the graph as an EPSC2 file:

```
print -depsc2 etopo1_filledcontourplot_vs1_matlab.eps
```

We can change the view of a surface plot using `view`.

```
figure1 = figure('Color',[1 1 1]);
axes1 = axes('Units','Centimeters',...
    'Position',[3 2 10 8],...
    'Box','on',...
    'View',[20 60],...
    'LineWidth',0.6,...
    'FontSize',8); hold on
surf1 = surf(LON,LAT,ETOPO1,...
    'FaceColor','interp',...
    'EdgeColor','none');
title1 = title('ETOPO1 Data Set',...
    'FontSize',10);
xlabel1 = xlabel('Longitude',...
    'FontSize',8);
```

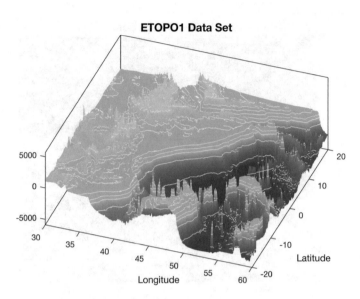

□ Fig. 6.3   Surface plot of the ETOPO1 elevation data, in combination with 3D contours. The function `demcmap` contained in the Mapping Toolbox creates and assigns a colormap appropriate for elevation data. The function `view` uses an azimuth of 20° and an elevation of 60° to change the view an observer sees of the 3D plot

```
ylabel1 = ylabel('Latitude',...
    'FontSize',8);
```

The command

```
print -depsc2 etopo1_surfaceplot_vs1_matlab.eps
```

exports the graph as an EPSC2 file. Once again, the result is obviously not a vector graph and we therefore export the graph as a JPEG file

```
print -djpeg70 -r600 etopo1_surfaceplot_vs2_matlab.jpg
```

compressed to 70%, with a resolution of 600 dpi. We can then combine the 3D graph with 3D contours stored in v (□Fig. 6.3).

```
figure1 = figure('Color',[1 1 1]);
axes1 = axes('Units','Centimeters',...
    'Position',[3 2 10 8],...
    'Box','on',...
    'View',[20 60],...
    'LineWidth',0.6,...
    'FontSize',8); hold on
surf1 = surf(LON,LAT,ETOPO1,...
    'FaceColor','interp',...
    'EdgeColor','none');
contour31 = contour3(LON,LAT,ETOPO1,v,...
    'Color',[1 1 1]);
title1 = title('ETOPO1 Data Set',...
    'FontSize',10);
xlabel1 = xlabel('Longitude',...
```

◨ Fig. 6.4 Surface plot of the ETOPO1 elevation data using `light`. The plot uses *Phong* as the light-ing type, which is very popular in 3D computer graphics, creating a combined diffuse and specular reflection on surfaces. The *specular exponent* of 20 defines the size of the highlight spot on the sur-face, which can vary between 0 and 500

```
    'FontSize',8);
ylabel1 = ylabel('Latitude',...
    'FontSize',8);
```

As before, the command

```
print -depsc2 etopo1_surfaceplotcontours_vs1_matlab.eps
```

exports the graph as an EPSC2 file. Yet again, the result is obviously not a vector graph and we therefore export the graph as a JPEG file

```
print -djpeg70 -r600 etopo1_surfaceplotcontours_vs2_matlab.jpg
```

compressed to 70%, with a resolution of 600 dpi. A spectacular way to plot digital topography is as a surface plot with angled lighting (◨Fig. 6.4).

```
figure1 = figure('Color',[1 1 1]);
axes1 = axes('Visible','off',...
    'Units','Centimeters',...
    'Position',[3 2 10 8],...
    'Box','on',...
    'View',[20 60],...
    'LineWidth',0.6,...
    'FontSize',8); hold on
surf1 = surf(LON,LAT,ETOPO1,...
    'SpecularExponent',20,...
    'FaceLighting','phong',...
    'FaceColor','interp',...
    'EdgeColor','none');
```

```
light1 = light('Style','local',...
    'Position',[145 70 155801]);
title1 = title('ETOPO1 Data Set',...
    'FontSize',10);
xlabel1 = xlabel('Longitude',...
    'FontSize',8);
ylabel1 = ylabel('Latitude',...
    'FontSize',8);
colormap1 = colormap(jet);
```

We again export the graph as a JPEG file

```
print -djpeg70 -r600 etopo1_surfaceplotlight_vs1_matlab.jpg
```

compressed to 70%, with a resolution of 600 dpi. Alternatively, we can export the graph, and all other previous graphs, as a PNG file

```
print -dpng -r600 etopo1_surfaceplotlight_vs1_matlab.png
```

with a resolution of 600 dpi.

6.4 The Global 30-Arc Second Elevation Data GTOPO30

The *Global 30 Arc Second Elevation Data* (GTOPO30) is a 30 arc second (approximately 1 km) global digital elevation data set that contains only elevation data, not bathymetry. The data set has been developed by the Earth Resources Observation System Data Center and is available from the web page

```
https://lta.cr.usgs.gov/GTOPO30
```

from where we are directed to the USGS EarthExplorer

```
http://earthexplorer.usgs.gov
```

The EarthExplorer can be used to search, preview, and download satellite images, aerial photographs and cartographic products through the US Geological Survey (USGS). The GTOPO30 data set is located under the Digital Elevation category. On this webpage we first select the desired map section in the *Search Criteria*, either by entering the coordinates of the four corners of the map or by zooming into the area of interest and selecting *Use Map*. As an example we enter the coordinates 2°8′37.58″N 36°33′47.06″E of the Suguta Valley in the Northern Kenya Rift. We then choose *GTOPO30* from the *Digital Elevation* collection as the *Data Set* and click *Results*. Clicking *Results* produces a list of records, together with a toolbar for previewing and downloading data. In our example we find a single data set GT30E020N40, acquired on 01 December 1996, centered at 15°N 40°E. We need to register with the USGS website, log on, and then download the data set as a 55.0 MB GeoTIFF and a 24.9 MB DEM, which is provided as a 26.1 MB compressed *zip* file. We can use the MATLAB unzip function to decompress the file *gt30e020n40_dem.zip*, as follows

```
clear, clc, close all

unzip('gt30e020n40_dem.zip')
```

After decompressing the file we obtain eight files containing the raw data and header files in various formats. The *zip* file also provides a GIF image of a shaded relief display of the data.

Importing the GTOPO30 data into the workspace is simple. The Mapping Toolbox provides an import routine `gtopo30` that reads the data and stores it onto a regular data grid. We import only a subset of the original matrix using the `gtopo30` command included in the Mapping Toolbox. Unfortunately, the GTOPO30 data have been renamed on the USGS EarthExplorer server and the function `gtopo30` has not yet been updated for this change. The binary file containing the digital elevation model extracted from *gt30e020n40_dem.zip* is named `gt30e020n40.dem` but `gtopo30` expects the file to be named `E020N40.dem`. We therefore rename the files by removing *gt30* from the file name in the unzipped collection and use

```
latlim = [-5 5]; lonlim = [30 40];
GTOPO30 = gtopo30('E020N40',1,latlim,lonlim);
```

to import the data. This script reads the data from the tile *e020n40* (without file extension) at full resolution (scale factor = 1) into the matrix `GTOPO30`, which has the dimensions of 1,200 by 1,200 cells. The coordinate system is defined by using the *lon/lat* limits listed above. The resolution is 30 arc seconds, corresponding to 1/120 of a degree.

```
[LON,LAT] = meshgrid(30:1/120:40-1/120,-5:1/120:5-1/120);
```

We need to reduce the limits by a factor of 120 in order to obtain a matrix of similar dimensions to the GTOPO30 matrix. A grayscale image can be generated from the elevation data using the function `surf`. The fourth power of the colormap `gray` is used to darken the map and the colormap is then flipped vertically in order to obtain dark colors for high elevations and light colors for low elevations, instead of the other way around.

```
figure1 = figure('Color',[1 1 1]);
axes1 = axes('Box','on',...
    'View',[0 90]); hold on
surf1 = surf(LON,LAT,GTOPO30,...
    'FaceColor','interp',...
    'EdgeColor','none');
xlabel1 = xlabel('Longitude');
ylabel1 = ylabel('Latitude');
colormap1 = colormap(flipud(gray.^4));
```

This script opens a new figure window and generates the gray surface using interpolated shading, displayed in an overhead view. We again display the graphs in a similar layout to that used in the previous chapter. The corresponding MATLAB script reads as follows:

```
v = 0 : 500 : 6000;
figure1 = figure('Color',[1 1 1]);
axes1 = axes('Units','Centimeters',...
    'Position',[3 2 10 8],...
    'Box','on',...
    'LineWidth',0.6,...
    'FontSize',8);
contourf1 = contourf(LON,LAT,GTOPO30,v);
```

⬛ Fig. 6.5   Surface plot of the GTOPO30 elevation data using `light`. The plot uses similar lighting settings to the previous plot, including the *Phong* lighting type and a *specular exponent* of 20

6

```
title1 = title('GTOPO30 Data Set',...
    'FontSize',10);
xlabel1 = xlabel('Longitude',...
    'FontSize',8);
ylabel1 = ylabel('Latitude',...
    'FontSize',8);
colormap1 = colormap(demcmap(GTOPO30));
```

The length of the *x*-axis is 10 cm, the length of the *y*-axis is 8 cm, the line thickness of the axes is 0.6 pt (0.6 points), the axis labels and title are in a sans serif font such as *Helvetica*, the font size of the axis labels is 8 pt and that of the title is 10 pt. The function `demcmap` contained in the Mapping Toolbox again creates and assigns a colormap appropriate for elevation data. The command

```
print -depsc2 gtopo30_filledcontourplot_vs1_matlab.eps
```

exports the graph as an EPSC2 file. We again plot the digital topography as a surface plot with angled lighting (⬛Fig. 6.5).

```
figure1 = figure('Color',[1 1 1]);
axes1 = axes('Visible','off',...
    'Units','centimeters',...
    'Position',[3 2 10 8],...
    'FontSize',8,...
    'View',[25 20],...
    'DataAspectRatio',[1 1 12000]); hold on
surf1 = surf(LON,LAT,SRTM_FILTERED,...f
    'SpecularExponent',20,...
    'FaceLighting','phong',...
    'FaceColor','interp',...
    'EdgeColor','none');
light1 = light('Parent',axes1,...
    'Style','local',...
    'Position',[145 70 155801]);
```

We again export the graph as a JPEG file

```
print -djpeg70 -r600 gtopo30_surfaceplotlight_vs1_matlab.jpg
```

compressed to 70%, with a resolution of 600 dpi.

6.5 The Shuttle Radar Topography Mission SRTM

The Shuttle Radar Topography Mission (SRTM) was a radar system on board the Space Shuttle *Endeavour* during an 11-day mission in February 2000 (Farr and Kobrick 2000; Farr et al. 2007). SRTM was an international project spearheaded by the National Geospatial-Intelligence Agency (NGA) and the National Aeronautics and Space Administration (NASA). Detailed information on the SRTM project, including a gallery of images and a user's forum, can be accessed through the NASA web page:

```
http://www2.jpl.nasa.gov/srtm/
```

The data were processed at the Jet Propulsion Laboratory. They are distributed through the United States Geological Survey's (USGS) National Map Viewer and Download Platform:

```
http://viewer.nationalmap.gov/launch/
```

Alternatively, the raw data files can be downloaded from

```
http://dds.cr.usgs.gov/srtm/
```

This directory contains zipped files of SRTM DEMs from various areas of the world, processed by the SRTM global processor and sampled at resolutions of 1 arc second (SRTM-1, 30 m grid) and 3 arc seconds (SRTM-3, 90 m grid). As an example we download the 1.7 MB file *s01e036.hgt.zip* from

```
http://dds.cr.usgs.gov/srtm/version2_1/SRTM3/Africa/
```

containing SRTM-3 data for the Kenya Rift Valley in East Africa. All elevations are in meters referenced to the WGS84 EGM96 geoid, as documented at

```
http://earth-info.nga.mil/GandG/wgs84/index.html
```

The name of this file refers to the longitude and latitude of the lower-left (southwest) pixel of the tile, i.e., latitude one degree south and longitude 36 degrees east SRTM-3 data contain 1,201 lines and 1,201 samples, with similar numbers of overlapping rows and columns. After having downloaded *S01E036.hgt.zip*, we use the MATLAB unzip function to decompress this file

```
clear, clc, close all

unzip('S01E036.hgt.zip')
```

and save *S01E036.hgt* in our working directory. The digital elevation model is provided as 16-bit signed integer data in a simple binary raster. The bit order is *big-endian* (Motorola's standard) with the most significant bit first. The data are imported into the workspace using

```
fid = fopen('S01E036.hgt','r');
SRTM = fread(fid,[1201,inf],'int16','b');
fclose(fid);
```

This script opens the file *s01e036.hgt* for read only access using `fopen` and defines the file identifier `fid`, which is then used to read the binaries from the file (using `fread`) and to write them into the matrix `SRTM`. Function `fclose` closes the file defined by `fid`. The matrix first needs to be transposed and flipped vertically.

```
SRTM = SRTM'; SRTM = flipud(SRTM);
```

The *−32768* flag for data voids can be replaced by `NaN`, which is the MATLAB representation for *Not-a-Number*.

```
SRTM(find(SRTM == -32768)) = NaN;
```

The SRTM data contain numerous gaps that might cause spurious effects during statistical analysis or when displaying the digital elevation model in a graph. A popular way to eliminate gaps in digital elevation models is by filling gaps with the arithmetic means of adjacent elements. We use the function nanmean since it treats `NaN`s as missing values and returns the mean of the remaining elements that are not `NaN`s. The following double for loop averages `SRTM(i-1:i+1,j-1:j+1)` arrays, i.e., averages over three-by-three element wide areas of the digital elevation model.

```
for i = 2 : 1200
    for j = 2 : 1200
        if isnan(SRTM(i,j))   == 1
            SRTM(i,j) = nanmean(nanmean(SRTM(i-1:i+1,j-1:j+1)));
        end
    end
end
clear i j
```

If there are still `NaN`s in the data set (as in our example) causing errors when importing the data set into a *Virtual Reality Modeling Language* (VRML) client, the double for loop can be run a second time. Finally, we check whether the data are now correctly stored in the workspace by printing the minimum and maximum elevations of the area.

```
max(SRTM(:))
min(SRTM(:))
```

In our example the maximum elevation of the area is 3,992 m above sea level and the minimum is 1,504 m. A coordinate system can be defined using the information that the lower-left corner is *s01e036*. The resolution is 3 arc seconds, corresponding to 1/1,200 of a degree.

```
[LON,LAT] = meshgrid(36:1/1200:37,-1:1/1200:0);
```

A shaded grayscale map can be generated from the elevation data using the function `surfl`. This function displays a shaded surface with simulated lighting.

```
figure1 = figure('Color',[1 1 1]);
axes1 = axes('Box','on',...
    'View',[0 90]); hold on
surf1 = surf(LON,LAT,SRTM,...
```

```
        'SpecularExponent',20,...
        'FaceLighting','phong',...
        'FaceColor','interp',...
        'EdgeColor','none');
   light1 = light('Style','local',...
        'Position',[145 70 155801]);
   xlabel1 = xlabel('Longitude');
   ylabel1 = ylabel('Latitude');
   colormap1 = colormap(flipud(gray));
```

This script opens a new figure window and generates the shaded-relief map using interpolated shading, as well as a gray colormap, displayed in an overhead view. Since SRTM data contain a large amount of noise, we first smooth the data using an arbitrary 9-by-9 pixel large moving average filter. The new matrix is then stored in the matrix SRTM_FILTERED.

```
   B = 1/81 * ones(9,9);
   SRTM_FILTERED = filter2(B,SRTM);
```

The corresponding shaded-relief map is generated by

```
   figure1 = figure('Color',[1 1 1]);
   axes1 = axes('Box','on',...
        'View',[0 90]); hold on
   surf1 = surf(LON,LAT,SRTM,...
        'SpecularExponent',20,...
        'FaceLighting','phong',...
        'FaceColor','interp',...
        'EdgeColor','none');
   light1 = light('Style','local',...
        'Position',[145 70 155801]);
   xlabel1 = xlabel('Longitude');
   ylabel1 = ylabel('Latitude');
   colormap1 = colormap(flipud(gray));
```

After having generated the shaded-relief map (◻Fig. 6.4), the plot must be exported to a graphics file. For instance, the figure may be written into a PNG format with 300 dpi resolution.

```
   print -dpng -r300 srtm_surfaceplotlight_vs1_matlab.png
```

The new file has a size of 621 KB; the decompressed image has a size of 16.5 MB. We again display the graphs in a similar layout to that used in the previous chapter. The corresponding MATLAB script reads as follows:

```
   figure1 = figure('Color',[1 1 1]);
   axes1 = axes('Units','Centimeters',...
        'Position',[3 2 10 8],...
        'Box','on',...
        'LineWidth',0.6,...
        'FontSize',8);
   surfl1 = surfl(LON,LAT,SRTM_FILTERED);
   title1 = title('SRTM Data Set',...
        'FontSize',10);
   xlabel1 = xlabel('Longitude',...
        'FontSize',8);
```

```
ylabel1 = ylabel('Latitude',...
    'FontSize',8);
shading interp
colormap gray
view(0,90)
```

The length of the *x*-axis is 10 cm, the length of the *y*-axis is 8 cm, the line thickness of the axes is 0.6 pt (0.6 points), the axis labels and title are in a sans serif font such as *Helvetica*, the font size of the axis labels is 8 pt and that of the title is 10 pt. We export the graph using the command

```
print -dpng -r600 srtm_surfaceplotlight_vs2_matlab.png
```

in a PNG format with 300 dpi resolution. We again plot the topography as a surface plot with angled light (◻Fig. 6.6).

6

```
figure1 = figure('Color',[1 1 1]);
axes1 = axes('Visible','off',...
    'Units','centimeters',...
    'Position',[3 2 10 8],...
    'FontSize',8,...
    'View',[25 20],...
    'DataAspectRatio',[1 1 12000]); hold on
surf1 = surf(LON,LAT,SRTM_FILTERED,...
    'SpecularExponent',20,...
    'FaceLighting','phong',...
    'FaceColor','interp',...
    'EdgeColor','none');
light1 = light('Parent',axes1,...
    'Style','local',...
    'Position',[145 70 155801]);
```

We again export the graph as a JPEG file

```
print -djpeg70 -r600 srtm_surfaceplotlight_vs3_matlab.jpg
```

compressed to 70%, with a resolution of 600 dpi.

◻ **Fig. 6.6**   Surface plot of the SRTM elevation data using `light`. The plot uses similar lighting set-tings to the previous plot, including the *Phong* lighting type and a *specular exponent* of 20

6.6 Interpolating and Visualizing Irregularly-Spaced Data

The previous data sets were all stored in evenly-spaced two-dimensional arrays. Most ground-based earth science data sets are, however, obtained from irregular sampling patterns. The data are therefore unevenly-spaced and need to be interpolated in order to allow a smooth and continuous surface to be computed from field measurements. MATLAB has, from the start, provided a biharmonic spline interpolation method that was developed by Sandwell (1987). This gridding method is particularly well suited for producing smooth surfaces from noisy data sets with unevenly-distributed control points. MATLAB today also provides other interpolation techniques such as bilinear, cubic spline, and nearest neighbor interpolation.

As an example we use synthetic *xyz* data representing the vertical distance between the surface of an imaginary stratigraphic horizon that has been displaced by a normal fault, and a reference surface. The foot wall of the fault shows roughly horizontal strata, whereas the hanging wall is characterized by the development of two large sedimentary basins. The *xyz* data are irregularly distributed and so need to be interpolated onto a regular grid. The *xyz* data are stored as a three-column table in a file named *normalfault.txt*.

```
4.3229698e+02    7.4641694e+01    9.7283620e-01
4.4610209e+02    7.2198697e+01    6.0655065e-01
4.5190255e+02    7.8713355e+01    1.4741054e+00
4.6617169e+02    8.7182410e+01    2.2842172e+00
4.6524362e+02    9.7361564e+01    1.1295175e-01
4.5526682e+02    1.1454397e+02    1.9007110e+00
4.2930233e+02    7.3175896e+01    3.3647807e+00
(cont'd)
```

The first and second columns contain the coordinates x (between 420 and 470 of an arbitrary spatial coordinate system) and y (between 70 and 120), while the third column contains the vertical z-values. The data are loaded using

```
clear, clc, close all

data = load('normalfault.txt');
```

Initially, we wish to create an overview plot of the spatial distribution of the control points. In order to label the points in the plot, numerical z-values of the third column are converted into character string representations with a maximum of two digits.

```
labels = num2str(data(:,3),2);
```

The 2D plot of our data is generated in two steps. Firstly, the data are displayed as empty circles using the `line` command. Secondly, the data are labeled using the function `text(x,y,'string')`, which adds text contained in `string` to the *xy* locations.

```
figure1 = figure('Color',[1 1 1]);
axes1 = axes('Box','on'); hold on
line1 = line(data(:,1),data(:,2),...
    'LineStyle','none',...
    'Marker','o',...
    'MarkerEdgeColor',[0.8 0.3 0.1]);
text1 = text(data(:,1),data(:,2),labels);
```

This plot helps us to define the axis limits for gridding and contouring: $xlim = [420\ 470]$ and $ylim = [70\ 120]$. The function `meshgrid` transforms the domain specified by vectors x and y into arrays XI and YI. The rows of the output array XI are copies of the vector x and the columns of the output array YI are copies of the vector y. We choose 1.0 as the grid interval.

```
x = 420:1:470; y = 70:1:120;
[XI,YI] = meshgrid(x,y);
```

The biharmonic spline interpolation is used to interpolate the irregular-spaced data at the grid points specified by XI and YI.

```
ZI = griddata(data(:,1),data(:,2),data(:,3),XI,YI,'v4');
```

The option `v4` selects the biharmonic spline interpolation, which was the sole gridding algorithm available until MATLAB4 was replaced by MATLAB5. MATLAB provides various tools with which to display the results. The simplest way to display the gridding results is as a contour plot using `contour`. By default, the number of contour levels and the values of the contour levels are chosen automatically. The choice of the contour levels depends on the minimum and maximum values of z.

```
figure1 = figure('Color',[1 1 1]);
axes1 = axes('Box','on'); hold on
contour1 = contour(XI,YI,ZI);
```

Alternatively, the number of contours can be chosen manually, e.g., ten contour levels.

```
figure1 = figure('Color',[1 1 1]);
axes1 = axes('Box','on'); hold on
contour1 = contour(XI,YI,ZI,10);
```

Contouring can also be performed at values specified in a vector v. Since the maximum and minimum values of z

```
min(data(:,3))
max(data(:,3))
```

are -27.4357 and 21.3018, we choose

```
v = -40 : 10 : 20;
```

The command

```
figure1 = figure('Color',[1 1 1]);
axes1 = axes('Box','on'); hold on
[cl,contour1] = contour(XI,YI,ZI,v);
```

yields contour matrix `cl` and a handle `contour1` that can be used as input to the function `clabel`, which labels contours automatically.

```
clabel(cl,contour1)
```

Alternatively, the plot can be labeled manually by selecting the `manual` option in the function `clabel`. This function places labels onto locations that have been selected with the mouse. Labeling is terminated by pressing the *return* key.

```
figure1 = figure('Color',[1 1 1]);
axes1 = axes('Box','on'); hold on
[c1,contour1] = contour(XI,YI,ZI,v);

clabel(c1,contour1,'manual')
```

Filled contours are an alternative to the empty contours used above. This function is used together with `colorbar`, which displays a legend for the plot. In addition, we can plot the locations (small circles) and *z*-values (contour labels) of the true data points (□Fig. 6.7).

```
figure1 = figure('Color',[1 1 1]);
axes1 = axes('Box','on'); hold on
contour1 = contourf(XI,YI,ZI,v);
line1 = line(data(:,1),data(:,2),...
    'LineStyle','none',...
    'Marker','o',...
    'MarkerEdgeColor',[0.8 0.3 0.1]);
text1 = text(data(:,1),data(:,2),labels);

print -depsc2 normalfault_filledcontourplot_vs1_matlab.eps
```

A pseudocolor plot is generated using the function `pcolor`. Black contours are also added at the same levels as in the above example.

```
figure1 = figure('Color',[1 1 1]);
axes1 = axes('Box','on'); hold on
```

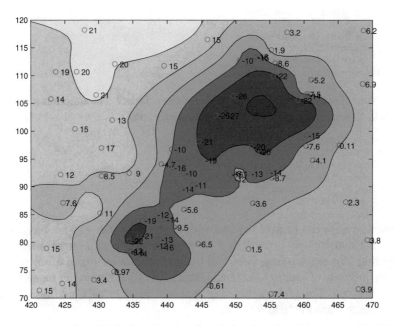

□ **Fig. 6.7** Contour plot with the locations (small circles) and *z*-values (contour labels) of the true data points

```
pcolor1 = pcolor(XI,YI,ZI);
pcolor1.LineStyle = 'none';
contour1 = contour(XI,YI,ZI,v,...
    'Color',[0 0 0]);
text1 = text(data(:,1),data(:,2),labels);
```

The third dimension is added to the plot using the `mesh` command. We can specify the direction of viewing by using the `View` property of `axes`, where the values `az`=−37.5 and `el`=30 define the default view for all 3D plots,

```
figure1 = figure('Color',[1 1 1]);
axes1 = axes('Box','on',...
    'View',[-37.5,30]); hold on
mesh1 = mesh(XI,YI,ZI);
```

whereas `az`=0 and `el`=90 is directly overhead and the default 2D view:

```
figure1 = figure('Color',[1 1 1]);
axes1 = axes('Box','on',...
    'View',[0,90]); hold on
mesh1 = mesh(XI,YI,ZI);
```

The function `mesh` provides one of many methods available in MATLAB for 3D presentation, another commonly used function being `surf`. The figure may be rotated by selecting the *Rotate 3D* option on the *Edit Tools* menu. We also introduce the function `colormap`, which uses predefined color look-up tables for 3D graphics. Typing `help graph3d` lists a number of built-in colormaps, although colormaps can also be arbitrarily modified and generated by the user. As an example we use the colormap `hot`, which is a *black-red-yellow-white* colormap.

```
figure1 = figure('Color',[1 1 1]);
axes1 = axes('Box','on',...
    'View',[-37.5,30]); hold on
surf1 = surf(XI,YI,ZI);
colormap1 = colormap(hot);
colorbar1 = colorbar;
```

Using *Rotate 3D* only rotates the 3D plot, not the colorbar. The function `surfc` combines both a surface and a 2D contour plot in one graph.

```
figure1 = figure('Color',[1 1 1]);
axes1 = axes('Box','on',...
    'View',[-37.5,30]); hold on
surf1 = surfc(XI,YI,ZI);
colormap1 = colormap(hot);
colorbar1 = colorbar;
```

The function `surf` can be used together with `light` to illustrate an advanced application for 3D visualization, generating a 3D colored surface with interpolated shading and lighting. The axis labeling, ticks, and background can be turned off by setting the `Visible` property of axes to `off`. In addition, black 3D contours can be added to the surface, as above. The grid resolution is increased prior to data plotting in order to obtain smooth surfaces (◨Fig. 6.8).

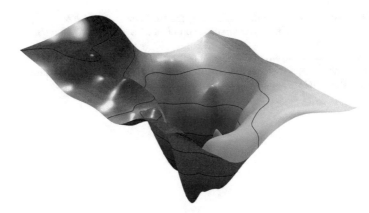

▣ Fig. 6.8 Three-dimensional surface with interpolated shading and simulated lighting. The axis labeling, ticks, and background are all turned off. The plot shows 3D contours as black lines

```
figure1 = figure('Color',[1 1 1]);
axes1 = axes('Box','on',...
    'View',[-37.5,30],...
    'Visible','off'); hold on
surf1 = surf(XI,YI,ZI,...
    'SpecularExponent',20,...
    'FaceLighting','phong',...
    'FaceColor','interp',...
    'EdgeColor','none');
light1 = light('Style','local',...
    'Position',[145 70 155801]);
contour31 = contour3(XI,YI,ZI,v,...
    'Color',[0 0 0]);
colormap1 = colormap(jet);
print -dpng -r600 normalfault_surfaceplotlight_vs1_matlab.png
```

The biharmonic spline interpolation described in this section provides a solution to most gridding problems. It was therefore, for some time, the only gridding method that came with MATLAB. However, different applications in earth sciences require different methods of interpolation, although they all have their problems.

Recommended Reading

Amante C, Eakins BW (2009) ETOPO1 1 arc-minute global relief model: procedures, data sources and analysis. NOAA Technical Memorandum NESDIS NGDC-24

Bohlander J, Scambos T (2007) Antarctic coastlines and grounding line derived from MODIS Mosaic of Antarctica (MOA). National Snow and Ice Data Center, Boulder, Colorado

Farr TG, Kobrick M (2000) Shuttle radar topography mission produces a wealth of data. American Geophysical Union Eos 81:583–585

Farr TG, Rosen P, Caro E, Crippen R, Duren R, Hensley S, Kobrick M, Paller M, Rodriguez E, Roth L, Seal D, Shaffer S, Shimada J, Umland J, Werner M, Oskin M, Burbank D, Alsdorf D (2007) The shuttle radar topography mission. Rev Geophysic 45:RG2004

Gorny AJ (1977) World Data Bank II general user guide Rep. PB 271869. Central Intelligence Agency, Washington DC

MathWorks (2016) Mapping toolbox—user's guide. The MathWorks, Natick, MA

Sandwell DT (1987) Biharmonic spline interpolation of GEOS-3 and SEASAT altimeter data. Geophys Res Lett 2:139–142

Soluri EA, Woodson VA (1990) World vector shoreline. Int Hydrogr Rev LXVII(1):27–35

Wessel P, Smith WHF (1996) A global self-consistent, hierarchical, high-resolution shoreline database. J Geophys Res 101(B4):8741–8743

6

Processing and Displaying Images in Earth Sciences

Electronic supplementary material The online version of this chapter (https://doi.org/10.1007/978-3-662-56203-1_7) contains supplementary material, which is available to authorized users.

7.1 **Introduction**

Computer graphics are stored and processed as either vector or raster data. Most of the data types that were encountered in the previous chapter were vector data, i.e., points, lines and polygons. Drainage networks, the outlines of geologic units, sampling locations, and topographic contours are all examples of vector data. In this chapter, coastlines are stored in a vector format while bathymetric and topographic data are saved in a raster format. Vector and raster data are often combined in a single data set, for instance to display the course of a river on a satellite image. Raster data are often converted to vector data by digitizing points, lines or polygons. Conversely, vector data are sometimes transformed to raster data.

Images are generally represented as raster data, i.e., as a 2D array of color intensities. Images are everywhere in geosciences. Field geologists use aerial photos and satellite images to identify lithologic units, tectonic structures, landslides and other features within a study area. Geomorphologists use such images to analyze drainage networks, river catchments, and vegetation or soil types. The analysis of images from thin sections, the automated identification of objects, and the measurement of varve thicknesses all make use of a great variety of image processing methods.

This chapter is concerned with the analysis and display of image data. The various ways that raster data can be stored on the computer are first explained (▶Sect. 7.2). The main tools for importing, manipulating and exporting image data are then presented in ▶Sect. 7.3. This knowledge is then used to process and to georeference satellite images (▶Sect. 7.4). On-screen digitization techniques are discussed in ▶Sect. 7.5. ▶Section 7.6 deals with image enhancement, correction and rectification. The Image Processing Toolbox is used for the specific examples throughout this chapter (Math-Works 2016). While the MATLAB User's Guide to the Image Processing Toolbox provides an excellent general introduction to the analysis of images, this chapter provides an overview of typical applications in earth sciences.

7.2 **Storing Images on a Computer**

Vector and raster graphics are the two fundamental methods for storing pictures. The typical format for storing *vector data* has already been introduced in the previous chapter. The *raster data* are stored as 2D arrays. The elements of these arrays represent variables such as the altitude of a grid point above sea level, the annual rainfall or, in the case of an image, the color intensity values.

```
174 177 180 182 182 182
165 169 170 168 168 170
171 174 173 168 167 170
184 186 183 177 174 176
191 192 190 185 181 181
189 190 190 188 186 183
```

Raster data can be visualized as 3D plots. The x and y figures are the indices of the 2D array or any other reference frame, and z is the numerical value of the elements of the array (see also ▶Chap. 6). The numerical values contained in the 2D array can be displayed as a pseudocolor plot, which is a rectangular array of cells with colors determined by a colormap. A colormap is an *m*-by-3 array of real numbers between 0.0 and 1.0. Each row defines a red, green, or blue (RGB) color. An example is the above array, which could be interpreted as grayscale intensities ranging from 0 (black) to 255 (white). More complex examples include satellite images that are stored in 3D arrays.

As previously discussed, a computer stores data as bits that have one of two states, represented by either a one or a zero (▶Chap. 4). If the elements of the 2D array represent the color intensity values of the *pixels* (short for *picture elements*) of an image, 1-bit arrays contain only ones and zeros.

```
0    0    1    1    1    1
1    1    0    0    1    1
1    1    1    1    0    0
1    1    1    1    0    1
0    0    0    0    0    0
0    0    0    0    0    0
```

This 2D array of ones and zeros can be simply interpreted as a black-and-white image, where the value of one represents white and zero corresponds to black. Alternatively, the 1-bit array could be used to store an image consisting of any two different colors, such as red and blue.

In order to store more complex types of data the bits are joined together to form larger groups, such as bytes consisting of eight bits. Since the earliest computers could only process eight bits at a time, early computer code was written in sets of eight bits, which came to be called bytes. Each element of the 2D array or pixel therefore contains a vector of eight ones or zeros.

```
1    0    1    0    0    0    0    1
```

These 8 bits (or 1 byte) allow $2^8 = 256$ possible combinations of the eight ones or zeros and are therefore able to represent 256 different intensities, such as grayscales. The 8 bits can be read in the following way, reading from right to left: a single bit represents two numbers, two bits represent four numbers, three bits represent eight numbers, and so forth up to a byte (or eight bits), which represents 256 numbers. Each added bit doubles the count of numbers. Here is a comparison of binary and decimal representations of the number 161:

```
128    64    32    16     8     4     2     1       (value of the bit)
  1     0     1     0     0     0     0     1       (binary)

128 +  0 + 32 +  0 +  0 +  0 +  0 +  1 = 161       (decimal)
```

The end members of the binary representation of grayscales are

```
0    0    0    0    0    0    0    0
```

which is black, and

```
1    1    1    1    1    1    1    1
```

which is pure white. In contrast to the above 1-bit array, the 1-byte array allows a grayscale image of 256 different levels to be stored. Alternatively, the 256 numbers could be interpreted as 256 discrete colors. In either case, the display of such an image requires an additional source of information concerning how the 256 intensity values are converted into colors. Numerous global colormaps for the interpretation of 8-bit color images exist that allow the cross-platform exchange of raster images, while local colormaps are often embedded in a graphics file.

The disadvantage of 8-bit color images is that the 256 discrete colorsteps are not enough to simulate smooth transitions for the human eye. A 24-bit system is therefore used in many applications, with 8 bits of data for each RGB channel giving a total of $256^3 = 16,777,216$ colors. Such a 24-bit image is stored in three 2D arrays, or one 3D array, of intensity values between 0 and 255.

```
195   189   203   217   217   221
218   209   187   192   204   206
207   219   212   198   188   190
203   205   202   202   191   201
190   192   193   191   184   190
186   179   178   182   180   169

209   203   217   232   232   236
234   225   203   208   220   220
224   235   229   214   204   205
223   222   222   219   208   216
209   212   213   211   203   206
206   199   199   203   201   187

174   168   182   199   199   203
198   189   167   172   184   185
188   199   193   178   168   172
186   186   185   183   174   185
177   177   178   176   171   177
179   171   168   170   170   163
```

Compared to the 1-bit and 8-bit representations of raster data, 24-bit storage certainly requires a lot more computer memory. In the case of very large data sets such as satellite images and digital elevation models the user should therefore think carefully about the most suitable way to store the data. The default data type in MATLAB is the 64-bit array, which allows storage of the sign of a number (bit 63), the exponent (bits 62–52) and roughly 16 significant decimal digits between approximately 10^{-308} and 10^{+308} (bits 51–0). However, MATLAB also works with other data types such as 1-bit, 8-bit and 24-bit raster data, to save memory.

The amount of memory required to store a raster image depends on the data type and the image's dimensions. The dimensions of an image can be described by the number of pixels, which is the number of rows of the 2D array multiplied by the number of columns. Let us assume an image of 729-by-713 pixels, such as the one we will use in the following section. If each pixel needs 8 bits to store a grayscale value, the memory required by the data is $729 \times 713 \times 8 = 4,158,216$ bits or $4,158,216/8 = 519,777$ bytes. This number is exactly what we obtain by typing `whos` in the command window. Common prefixes for bytes are kilo-, mega-, giga- and so forth.

```
bit = 1 or 0 (b)
8 bits = 1 byte (B)
1024 bytes = 1 kilobyte (KB)
1024 kilobytes = 1 megabyte (MB)
1024 megabytes = 1 gigabyte (GB)
1024 gigabytes = 1 terabyte (TB)
```

Note that in data communication 1 kilobit = 1,000 bits, while in data storage 1 kilobyte = 1,024 bytes. A 24-bit or *true color image* then requires three times the memory required to store an 8-bit image, or 1,559,331 bytes = 1,559,331/1,024 kilobytes (KB) ≈ 1,523 KB ≈ 1,559,331/1,024^2 = 1.487 megabytes (MB).

However, the dimensions of an image are often given, not by the total number of pixels, but by the length and height of the image and its resolution. The resolution of an image is the number of *pixels per inch* (ppi) or *dots per inch* (dpi). The standard resolution of a computer monitor is 72 dpi although modern monitors often have a higher resolution such as 96 dpi. For instance, a 17 inch monitor with 72 dpi resolution displays 1,024-by-768 pixels. If the monitor is used to display images at a different (lower, higher) resolution, the image is resampled to match the monitor's resolution. For scanning and printing, a resolution of 300 or 600 dpi is enough in most applications. However, scanned images are often scaled for large printouts and therefore have higher resolutions such as 2,400 dpi. The image used in the next section has a width of 25.2 cm (or 9.92 inches) and a height of 25.7 cm (10.12 inches). The resolution of the image is 72 dpi. The total number of pixels is therefore 72 × 9.92 ≈ 713 in a horizontal direction, and 72 × 10.12 ≈ 729 in a vertical direction.

7.3 Importing, Processing and Exporting Images

We first need to learn how to read an image from a graphics file into the workspace. As an example we use a satellite image showing a 10.5 by 11 km subarea in northern Chile:

```
http://asterweb.jpl.nasa.gov/gallery/images/unconform.jpg
```

The file *unconform.jpg* is a processed TERRA-ASTER satellite image that can be downloaded free-of-charge from the NASA web page. We save this image in the working directory. The command

```
clear, clc, close all

I1 = imread('unconform_image_vs1_original.jpg');
```

reads and decompresses the JPEG file, imports the data as a 24-bit RGB image array and stores it in a variable I1. The command

```
whos
```

shows how the RGB array is stored in the workspace:

```
Name      Size            Bytes  Class     Attributes
I1        729x713x3     1559331  uint8
```

The details indicate that the image is stored as a 729-by-713-by-3 array, representing a 729-by-713 array for each of the colors red, green and blue. The listing of the current variables in the workspace also gives the information *uint8* array, i.e., each array element representing one pixel contains 8-bit integers. These integers represent intensity values between 0 (minimum intensity) and 255 (maximum). As an example, here is a sector in the upper-left corner of the data array for red:

```
I1(50:55,50:55,1)

ans =
    174 177 180 182 182 182
    165 169 170 168 168 170
    171 174 173 168 167 170
    184 186 183 177 174 176
    191 192 190 185 181 181
    189 190 190 188 186 183
```

We can now view the image using the command

```
imshow(I1)
```

which opens a new Figure Window showing an RGB composite of the image. In contrast to the RGB image, a grayscale image needs only a single array to store all the necessary information. We therefore convert the RGB image into a grayscale image using the command `rgb2gray` (RGB to gray):

```
I2  =  rgb2gray(I1);
```

The new workspace listing now reads

```
Name        Size            Bytes  Class    Attributes
I1          729x713x3     1559331  uint8
I2          729x713        519777  uint8
ans         6x6                36  uint8
```

in which the difference between the 24-bit RGB and the 8-bit grayscale arrays can be observed. The variable `ans` for *Most recent answer* was created above using `I1(50:55,50:55,1)`, without assigning the output to another variable. The command

```
imshow(I2)
```

displays the result (◘Fig. 7.1). It is easy to see the difference between the two images in separate Figure Windows. Let us now process the grayscale image. First, we compute a histogram of the distribution of intensity values.

```
imhist(I2)
```

A simple technique to enhance the contrast in such an image is to transform this histogram to obtain an equal distribution of grayscales.

```
I3 = histeq(I2);
```

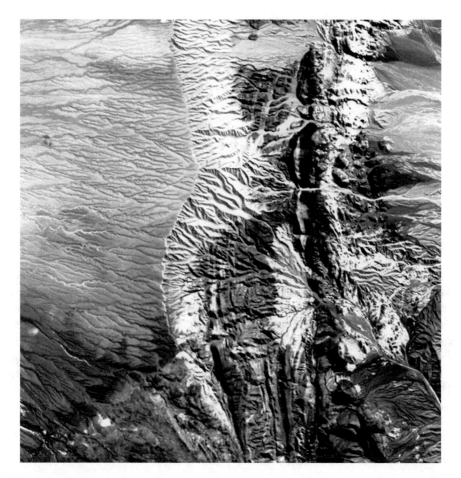

◨ Fig. 7.1 Grayscale image. After converting the RGB image stored in a 729 × 713 × 3 array into a grayscale image stored in a 729 × 713 array, the result is displayed using imshow. Original image courtesy of NASA/GSFC/METI/ERSDAC/JAROS and U.S./Japan ASTER Science Team

We can view the difference again using

```
imshow(I3)
```

and save the results in a new file.

```
imwrite(I3,'unconform_image_vs2_matlab.jpg')
```

We can read the header of the new file by typing

```
imfinfo('unconform_image_vs2_matlab.jpg')
```

which yields

```
ans =
                Filename:
'/Users/trauth/Desktop/matlab/unconform_image_vs2_matlab.jpg'
            FileModDate: '09-Sep-2016 16:25:20'
               FileSize: 138419
                 Format: 'jpg'
          FormatVersion: ''
                  Width: 713
                 Height: 729
               BitDepth: 8
              ColorType: 'grayscale'
        FormatSignature: ''
        NumberOfSamples: 1
           CodingMethod: 'Huffman'
          odingProcess: 'Sequential'
                Comment: {}
```

Hence, the command `imfinfo` can be used to obtain useful information (name, size, format, and color type) concerning the newly-created image file.

There are many ways of transforming the original satellite image into a practical file format. The image data could, for instance, be stored as an *indexed color image*, which consists of two parts: a colormap array and a data array. The colormap array is an m-by-3 array containing floating-point values between 0 and 1. Each column specifies the intensity of the red, green and blue colors. The data array is an x-by-y array containing integer elements corresponding to the lines m of the colormap array, i.e., the specific RGB representation of a certain color. Let us transfer the above RGB image into an indexed image. The colormap of the image should contain 16 different colors. The result of

```
[I4,map] = rgb2ind(I1,16);
imshow(I1), figure, imshow(I4,map)
```

saved as another JPEG file using

```
imwrite(I4,map,'unconform_image_vs3_matlab.jpg')
```

clearly shows the difference between the original 24-bit RGB image (256^3 or about 16.7 million different colors) and a color image of only 16 different colors. The display of the image uses the default colormap of MATLAB. Typing

```
imshow(I4,map)
cmap = colormap
```

actually retrieves the 16-by-3 array of the current colormap

```
cmap =
    0.0588    0.0275    0.0745
    0.5490    0.5255    0.4588
    0.7373    0.7922    0.7020
    0.3216    0.2706    0.2667
    0.6471    0.6784    0.6157
    0.7961    0.8549    0.9176
    0.4510    0.3922    0.3333
    0.2000    0.1451    0.1451
    0.4824    0.5412    0.5843
```

```
0.4039    0.4078    0.4784
0.6667    0.7020    0.7451
0.8980    0.8745    0.7255
0.2824    0.2902    0.4039
0.9569    0.9647    0.9608
0.1765    0.1686    0.2902
0.5843    0.5843    0.6078
```

We can replace the default colormap by any other built-in colormap. Typing

```
help graph3d
```

lists the available colormaps. As an example we can use

```
imshow(I4,map)
colormap(hot)
```

to display the image with a black-red-yellow-white colormap. Typing

```
edit hot
```

reveals that `hot` is a function creating the *m*-by-3 array containing floating-point values between 0 and 1. We can also design our own colormaps, either by manually creating an *m*-by-3 array or by creating another function similar to `hot`. Alternatively, we can also use random numbers

```
rng(0)
map = rand(16,3);
imshow(I4,map)
```

to display the image with random colors. Finally, we can create an indexed color image of three different colors, displayed with a simple colormap of full intensity red, green and blue.

```
[I5,map] = rgb2ind(I1,3);
imshow(I5,[1 0 0;0 1 0;0 0 1])
```

Typing

```
imwrite(I4,map,'unconform_image_vs4_matlab.jpg')
```

saves the result as another JPEG file.

7.4 Processing and Georeferencing Satellite Images

In ▶Sect. 7.3 we used a processed ASTER image that we downloaded from the ASTER webpage. In this section we will use raw data from this sensor. The ASTER sensor is mounted on the TERRA satellite launched in 1999, part of the Earth Observing System (EOS) series of multi-national NASA satellites (Abrams and Hook 2002). ASTER stands for *Advanced Spaceborne Thermal Emission and Reflection Radiometer*, providing high-resolution (15–90 m) images of the earth in 14 bands, including three visible to near infrared bands (VNIR bands 1–3), six short-wave infrared bands (SWIR bands 4–9), and five thermal (or long-wave) infrared bands (TIR bands 10–14). ASTER images are used to map the surface temperature, emissivity, and reflectance of the earth's surface. The 3rd near infrared band is recorded twice: once with the sensor pointing directly downwards (band 3N, where *N* stands for *nadir* from the Arabic word for *opposite*), as it does for all other channels, and a second time with the sensor angled backwards at 27.6° (band 3B, where *B* stands for backward looking). These two bands are used to generate ASTER digital elevation models (DEMs).

The ASTER instrument produces two types of data: Level-1A (L1A) and Level-1B (L1B) data (Abrams and Hook 2002). Whereas the L1A data are reconstructed, unprocessed instrument data, the L1B data are radiometrically and geometrically corrected. Since 1 April 2016 all ASTER data has been available at no charge, through USGS' GloVis

```
http://glovis.usgs.gov
```

or the NASA's Earthdata Search Client

```
https://search.earthdata.nasa.gov/search
```

Visit the *Land Processes Distributed Active Archive Center* (LP DAAC) website

```
https://lpdaac.usgs.gov/dataset_discovery/aster
```

to learn more about ASTER. As an example we process an image from an area in Kenya showing Lake Naivasha. On NASA's Earthdata Search Client webpage we first define the desired location by zooming into the area of interest, which in our example is the Central Kenya Rift northwest of Nairobi, and then use the *Search by spatial point* icon on the right and click on the map to place a location marker. As an example, we enter the coordinates of 0° 46′ 31.38″S 36° 22′ 17.31″E for Lake Naivasha in the Central Kenya Rift. We next select *Aster* from the *Instruments* menu on the left of the webpage. We obtain a list of 17 appropriate granules where we choose *ASTER L1A Reconstructed Unprocessed Instrument Data V003*. We get a list of 228 items that we sort using *Start Data, Oldest first*. We search the previews of images in the list for cloud-free images with Lake Naivasha in the center. As an example, we use the image with the name *AST_L1A_003_03082003080706_03242003202838* acquired on 8 March 2003, which we can download using the *Retrieve single granule data* button next to the *View granule details* button in the list. We have to create an account and login, after which we are directed to a webpage entitled *Review and Select Service Options*. In our example we see that we have selected *1 Granule* with a size of 116.3 MB. At the bottom of the page we change the *Data Format* to HDF-EOS and then click *Continue*. The next page shows a summary of our contact information, where we click Submit to download the data. The data are then processed by the server and, according to the webpage, an email containing download links will be sent to the address that we provided. After a few minutes we receive this email providing a download link. The Level-1A data are stored in two files:

```
AST_L1A_003_03082003080706_03242003202838.hdf
AST_L1A_003_03082003080706_03242003202838.hdf.met
```

The first file (116 MB) contains the actual raw data while the second file (102 KB) contains the header, together with all sorts of information about the data. We save both files in our working directory. Since the file name is very long, we first save it in the `filename` variable and then use `filename` instead of the long file name. We then need to modify only this single line of MATLAB code if we want to import and process other satellite images.

```
filename = 'AST_L1A_003_03082003080706_03242003202838.hdf';
```

The Image Processing Toolbox contains various tools for importing and processing files stored in the hierarchical data format (HDF). The graphical user interface (GUI) based import tool for importing certain parts of the raw data is

```
hdftool('filename')
```

This command opens a GUI that allows us to browse the content of the HDF file, obtains all information on the contents, and imports certain frequency bands of the satellite image. Alternatively, the command `hdfread` can be used as a quicker way of accessing image data. The *vnir_Band3n*, *vnir_Band2*, and *vnir_Band1* typically contain much information about lithology (including soils), vegetation and water on the earth's surface. These bands are therefore usually combined into 24-bit RGB images. We first read the data

```
I1 = hdfread(filename,'VNIR_Band3N','Fields','ImageData');
I2 = hdfread(filename,'VNIR_Band2','Fields','ImageData');
I3 = hdfread(filename,'VNIR_Band1','Fields','ImageData');
```

These commands generate three 8-bit image arrays, each representing the intensity within a certain infrared (IR) frequency band of a 4200-by-4100 pixel image. We are not using the data for quantitative analyses and therefore do not need to convert the digital number (DN) values into radiance and reflectance values. The *ASTER User Handbook* by Abrams and Hook (2002) provides the necessary information on these conversions. We will instead process the ASTER image to create a georeferenced RGB composite of bands 3N, 2 and 1, for use in fieldwork. We first use a contrast-limited adaptive histogram equalization method to enhance the contrast in the image by typing

```
I1 = adapthisteq(I1);
I2 = adapthisteq(I2);
I3 = adapthisteq(I3);
```

and then concatenate the result to a 24-bit RGB image using `cat`.

```
naivasha_rgb = cat(3,I1,I2,I3);
```

As with the previous examples, the 4200-by-4100-by-3 array can now be displayed using

```
imshow(naivasha_rgb,'InitialMagnification',10)
```

setting the initial magnification of this very large image to 10% (❑Fig. 7.2). MATLAB scales images to fit the computer screen. Exporting the processed image from the Figure Window, we only save the image at the monitor's resolution. To obtain an image at a higher resolution, we use the command

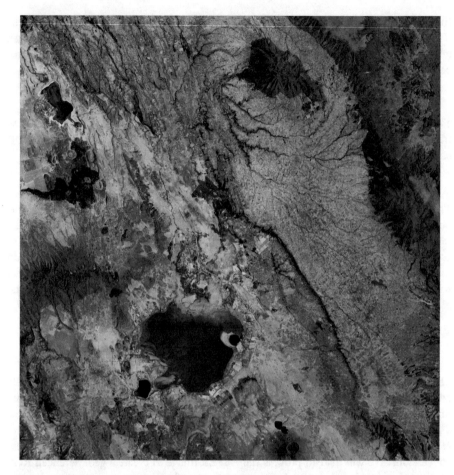

⬛ Fig. 7.2 RGB composite of a TERRA-ASTER image using the spectral infrared bands *vnir_Band3n*, *2* and *1*. The result is displayed using `imshow`. Original image courtesy of NASA/GSFC/METI/ERSDAC/ JAROS and U.S./Japan ASTER Science Team

```
imwrite(naivasha_rgb,'naivasha_image_vs1_matlab.tif','tif')
```

This command saves the RGB composite as a 52 MB TIFF file *naivasha.tif* in the working directory, which can then be processed using other software such as Adobe Photoshop. The processed ASTER image does not yet have a coordinate system and therefore needs to be tied to a geographical reference frame (*georeferencing*). The HDF browser

```
hdftool('filename')
```

can be used to extract the geodetic coordinates of the four corners of the image. This information is contained in the header of the HDF file. Having launched the HDF tool, we select the uppermost directory called *naivasha.hdf* and find a long list of file attributes in the upper right panel of the GUI, one of which is *productmetadata.0,* which includes the attribute *scenefourcorners*. We collect the coordinates of the four scene corners into a single array inputpoints:

```
inputpoints(1,:) = [36.214332 -0.319922];   % upper left corner
inputpoints(2,:) = [36.096003 -0.878267];   % lower left corner
inputpoints(3,:) = [36.770406 -0.400443];   % upper right corner
inputpoints(4,:) = [36.652213 -0.958743];   % lower right corner
```

It is important to note that the coordinates contained in *productmetadata.0* need to be flipped in order to have *x = longitudes* and *y = latitudes*. The four corners of the image correspond to the pixels in the four corners of the image, which we store in a variable named basepoints.

```
basepoints(1,:) = [1,1];               % upper left pixel
basepoints(2,:) = [1,4200];            % lower left pixel
basepoints(3,:) = [4100,1];            % upper right pixel
basepoints(4,:) = [4100,4200];         % lower right pixel
```

The function fitgeotrans now takes the pairs of control points, inputpoints and basepoints, and uses them to infer a spatial transformation matrix tform.

```
tform = fitgeotrans(inputpoints,basepoints,'affine');
```

We next determine the limits of the input (i.e., of the original image naivasha_rgb) for georeferencing using size, which yields xLimitsIn and yLimitsIn. Adding a value of 0.5 to both xLimitsIn and yLimitsIn prevents the edges of the image from being truncated during the affine transformation. We then determine the limits of the output (i.e. of the georeferenced image, which is subsequently called newnaivasha_rgb) using outputLimits, which yields XBounds and YBounds.

```
xLimitsIn = 0.5 + [0 size(naivasha_rgb,2)];
yLimitsIn = 0.5 + [0 size(naivasha_rgb,1)];
[XBounds,YBounds] = outputLimits(tform,xLimitsIn,yLimitsIn);
```

We then use imref2d to reference the image to a world (or global) coordinate system.

```
Rout = imref2d(size(naivasha_rgb),XBounds,YBounds);
```

An imref2d object encapsulates the relationship between the intrinsic coordinates anchored to the rows and columns of the image, and the spatial location of the same row and column locations within a world coordinate system. Finally, the affine transformation can be applied to the original RGB composite naivasha_rgb in order to obtain a georeferenced version of the satellite image newnaivasha_rgb with the same size as naivasha_rgb.

```
newnaivasha_rgb = imwarp(naivasha_rgb,tform,'OutputView',Rout);
```

An appropriate grid for the image can now be computed. The grid is typically defined by the minimum and maximum values for the longitude and latitude. The vector increments are then obtained by dividing the ranges of the longitude and latitude by the array's dimensions and then subtracting one from the results. Note the difference

7

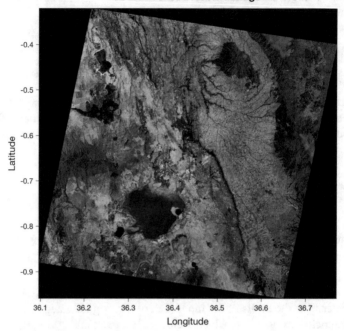

□ Fig. 7.3   Geoferenced RGB composite of a TERRA-ASTER image using the infrared bands *vnir_Band3n*, *2* and *1*. The result is displayed using imshow. Original image courtesy of NASA/GSFC/METI/ERSDAC/ JAROS and U.S./Japan ASTER Science Team

between the MATLAB numbering convention and the common coding of maps used in published literature. The north/south suffix is generally replaced by a negative sign for south, whereas MATLAB coding conventions require negative signs for north.

```
X = 36.096003 : (36.770406 - 36.096003)/4100 : 36.770406;
Y = -0.958743 : ( 0.958743 -  0.319922)/4200 : -0.319922;
```

The georeferenced image is displayed with coordinates on the axes and a superimposed grid (□Fig. 7.3). By default, the function imshow inverts the latitude axis when images are displayed by setting the YDir property to Reverse. To invert the latitude axis direction back to normal, we need to set the YDir property to Normal by typing

```
figure1 = figure('Color',[1 1 1]);
axes1 = axes; hold on
imshow1 = imshow(newnaivasha_rgb,...
   'XData',X,...
   'YData',Y,...
   'InitialMagnification',10);
axes1.YDir = 'normal';
axes1.Visible = 'on';
axis tight
xlabel1 = xlabel('Longitude');
ylabel1 = ylabel('Latitude');
title1 = title('Georeferenced ASTER Image');
```

Exporting the image is possible in many different ways, for example using

```
print -dpng -r600 naivasha_georef_vs1_matlab.png
```

to export it as a PNG file with a resolution of 600 dpi.

In the previous example we used the geodetic coordinates of the four corners to georeference the ASTER image. The Image Processing Toolbox also includes functions to automatically align two images that are shifted and/or rotated with respect to each other, cover slightly different areas, or have different resolutions. We use two ASTER images of the Suguta Valley in the Northern Kenya Rift as an example. The images have been processed in the same way as described for the image of Lake Naivasha and exported as TIFF files using `imwrite`. The image in the file *sugutavalley_image_1_vs1_original.tif* was taken on 20th February 2003 and the second image in *sugutavalley_image_2_vs1_original* was taken on 31st August 2003, both just after 8 o'clock in the morning. Lake Logipi, in the center of the images, is much larger in the second image than in the first image. The original images are otherwise almost identical, except for the second image being shifted slightly towards the east. To demonstrate the automatic alignment of the images, the second image has been rotated counterclockwise by five degrees. Furthermore, both images have been cropped to the area of the Suguta Valley, including a small section of the rift shoulders to the west and to the east. We import both images using

```
clear, clc, close all

image1 = imread('sugutavalley_image_1_vs1_original.tif');
image2 = imread('sugutavalley_image_2_vs1_original.tif');
```

The size of `image1` is 666-by-329-by-3, while `image2` is slightly smaller: 614-by-270-by-3. We display the images by typing

```
subplot(1,2,1), imshow(image1)
subplot(1,2,2), imshow(image2)
```

The function `imregconfig` creates optimizer and metric configurations that we transfer into `imregister` to perform intensity-based image registration,

```
[optimizer, metric] = imregconfig('monomodal');
```

where `monomodal` assumes that the images were captured by the same sensor. We can use these configurations to calculate the spatial transformation matrix `tform` using the transformation type `affine`, as in the previous example.

```
tform = imregtform(image2(:,:,1),image1(:,:,1), ...
    'affine',optimizer,metric);
```

This transformation can be applied to `image2` in order to automatically align it with `image1`.

```
image2_reg = imwarp(image2,tform,'OutputView', ...
    imref2d(size(image1)));
```

(a) **(b)** **(c)**

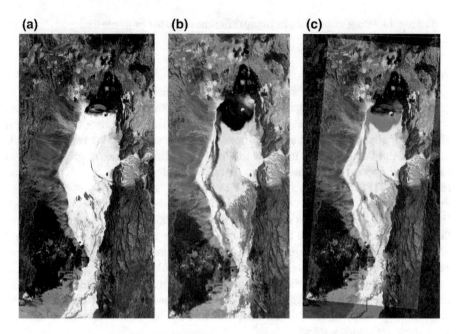

⬛ Fig. 7.4 Automatically aligned TERRA-ASTER images of the Suguta Valley in the Northern Kenya Rift, **a** first image taken on 20th February 2003, **b** second image taken on 31st August 2003, and **c** overlay of the second image aligned with the first image. Original image courtesy of NASA/GSFC/METI/ERSDAC/JAROS and U.S./Japan ASTER Science Team

We can compare the result with the original images using

```
subplot(1,3,1), imshow(image1)
subplot(1,3,2), imshow(image2)
subplot(1,3,3), imshowpair(image1,image2_reg,'blend')
print -dpng -r600 sugutavalley_aligned_vs1_matlab.png
```

As we can see, the second image is now nicely aligned with the first image (⬛Fig. 7.4). The two images can now be used to map changes in the area (e.g., in the size of the lake) between 20th February and 31st August 2003. This script can also be used to automatically align other images, in particular those captured by different sensors.

7.5 Digitizing from the Screen: From Pixel to Vector

On-screen digitizing is a widely-used image processing technique. While practical digitizer tablets exist in all formats and sizes, most people prefer digitizing vector data from the screen. Examples of this type of application include the digitizing of river networks and catchment areas on topographic maps, of the outlines of lithologic units on geological maps, of landslide distributions on satellite images, and of mineral grain distributions in microscope images. The digitizing procedure consists of the following steps. The image is first imported into the workspace. A coordinate system is then defined, allowing the objects of interest to be entered by moving a cursor or cross hair onto it and clicking the mouse button. The result is a two-dimensional array of xy data comprising, for example, longitudes and latitudes of the corner points of a polygon, or the coordinates of the objects of interest in a particular area.

The function `ginput` included in the standard MATLAB toolbox allows graphical input from the screen, using a mouse. It is generally used to select data points from a figure created by an arbitrary graphics function such as `plot`. The function `ginput` is often used for interactive plotting, i.e., the digitized points appearing on the screen after they have been selected. The disadvantage of the function is that it does not provide coordinate referencing on an image. We therefore use a modified version of the function, which allows an image to be referenced to an arbitrary rectangular coordinate system. Save the following code for this modified version of the function `ginput` in a text file `minput.m`.

```
function data = minput(imagefile)
% Specify the limits of the image
xmin = input('Specify xmin! ');
xmax = input('Specify xmax! ');
ymin = input('Specify ymin! ');
ymax = input('Specify ymax! ');

% Read image and display
B = imread(imagefile);
a = size(B,2); b = size(B,1);
imshow(B);

% Define lower left and upper right corner of image
disp('Click on lower left and upper right corner, then  < return > ')
[xcr,ycr]   = ginput;
XMIN = xmin-((xmax-xmin)*xcr(1,1)/(xcr(2,1)-xcr(1,1)));
XMAX = xmax+((xmax-xmin)*(a-xcr(2,1))/(xcr(2,1)-xcr(1,1)));
YMIN = ymin-((ymax-ymin)*ycr(1,1)/(ycr(2,1)-ycr(1,1)));
YMAX = ymax+ ((ymax-ymin)*(b-ycr(2,1))/(ycr(2,1)-ycr(1,1)));

% Digitize data points
disp('Click on data points to digitize, then < return > ')
[xdata,ydata] = ginput;
XDATA =  XMIN + ((XMAX-XMIN)*xdata/size(B,2));
YDATA =  YMIN + ((YMAX-YMIN)*ydata/size(B,1));
data(:,1) = XDATA; data(:,2) = YDATA;
```

The function `minput` has four stages. In the first stage the user enters the limits of the coordinate axes as reference points for the image. The image is then imported into the workspace and displayed on the screen. The third stage uses `ginput` to define the upper left and lower right corners of the image. In the fourth stage the relationship between the coordinates of the two corners on the figure window and the reference coordinate system is then used to compute the transformation for all of the digitized points.

As an example we use the image stored in the file *naivasha_georef_vs1_matlab.jpg* and digitize the outline of Lake Naivasha in the center of the image. We activate the new function `minput` from the Command Window using the commands

```
clear, clc, close all

data = minput('naivasha_georef_vs1_matlab.png');
```

The function first asks for the coordinates of the limits of the *x*-axis and the *y*-axis, for the reference frame. We enter the corresponding numbers and press *return* after each input.

```
Specify xmin! 36.1
Specify xmax! 36.7
Specify ymin! -1
Specify ymax! -0.3
```

The function then reads the file *naivasha_georef_vs1_matlab.jpg* and displays the image. We ignore the warning

```
Warning: Image is too big to fit on screen; displaying at 33%
```

and wait for the next response

```
Click on lower left and upper right corner, then <return>
```

The image window can be scaled according to user preference. Clicking on the lower left and upper right corners defines the dimensions of the image. These changes are registered by pressing *return*. The routine then references the image to the coordinate system and waits for the input of the points we wish to digitize from the image.

```
Click on data points to digitize, then <return>
```

We finish the input by again pressing *return*. The *xy* coordinates of our digitized points are now stored in the variable `data`. We can now use these vector data for other applications.

7.6   Image Enhancement, Correction and Rectification

This section introduces some fundamental tools for image enhancement, correction and rectification. As an example we use an image of varved sediments deposited around 33 kyrs ago in a landslide-dammed lake in the Quebrada de Cafayate of Argentina (25° 58.900′S 65° 45.676′W) (Trauth et al. 1999, 2003). The diapositive was taken on 1st October 1996 with a film-based single-lens reflex (SLR) camera. A 30-by-20 cm print was made from the slide, which has been scanned using a flatbed scanner and saved as a 394 KB JPEG file. We use this as an example because it demonstrates some problems that we can solve with the help of image enhancement (❒Fig. 7.5).

We can read and decompress the file *varves_image_vs1_original.jpg* by typing

```
clear, clc, close all

I1 = imread('varves_image_vs1_original.jpg');
```

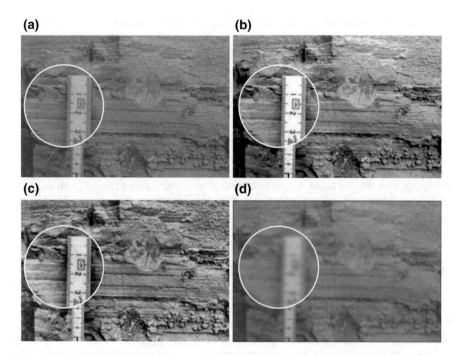

☐ Fig. 7.5 Results of image enhancements, **a** original image, **b** image with intensity values adjusted using imadjust, Gamma = 1.5, **c** image with contrast enhanced using adapthisteq, **d** image after filtering with a 20-by-20 pixel filter with the shape of a Gaussian probability density function with a mean of zero and a standard deviation of 10, using fspecial and imfilter

which yields a 24-bit RGB image array `I1` in the MATLAB workspace. Typing

```
whos
```

yields

```
Name          Size                   Bytes   Class       Attributes
I1            1096x1674x3            5504112  uint8
```

revealing that the image is stored as a `uint8` array with a the size of 1,096-by-1,674-by-3, with 1,096-by-1,674 arrays for each color (red, green and blue). We can display the image using the command

```
imshow(I1)
```

which opens a new Figure Window showing an RGB composite of the image. As we see, the image has a low level of contrast and very pale colors, and the sediment layers are not exactly horizontal. These are characteristics of the image that we want to improve in the following steps.

We first adjust the image intensity values or colormap. The function `imadjust(I1,[li; hi],[lo ho])` maps the values of the image `I1` to new values in I2, such that values between li and hi are adjusted to values between lo and ho. Values below li and above hi are clipped, i.e., these values are adjusted to lo and ho, respectively. We can determine the range of the pixel values using

```
lh = stretchlim(I1)
```

which yields

```
lh =
    0.3255    0.2627    0.2784
    0.7020    0.7216    0.7020
```

indicating that the red color ranges from 0.3255 to 0.7020, green ranges from 0.2627 to 0.7216, and blue ranges from 0.2784 to 0.7020. We can utilize this information to automatically adjust the image with imadjust by typing

```
I2 = imadjust(I1,lh,[]);
```

which adjusts the ranges to the full range of [0, 1], and then display the result.

```
imshow(I2)
```

We can clearly see the difference between the very pale image I1 and the more saturated image I2. The parameter gamma in imadjust(I1,[li;hi],[lo;ho],gamma) specifies the shape of the curve describing the relationship between I1 and I2. If gamma<1 the mapping is weighted toward higher (brighter) output values. If gamma>1 the mapping is weighted toward lower (darker) output values. The default value of gamma=1 causes linear mapping of the values in I1 to new values in I2.

```
I3 = imadjust(I1,lh,[],0.5);
I4 = imadjust(I1,lh,[],1.5);

subplot(2,2,1), imshow(I1)
    title('Original Image')
subplot(2,2,2), imshow(I2)
    title('Adjusted Image, Gamma=1.0')
subplot(2,2,3), imshow(I3)
    title('Adjusted Image, Gamma=0.5')
subplot(2,2,4), imshow(I4)
    title('Adjusted Image, Gamma=1.5')
```

We can use imhist to display a histogram showing the distribution of intensity values for the image. Since imhist only works for two-dimensional images, we examine the histogram of the red color only.

```
subplot(2,2,1), imhist(I1(:,:,1))
    title('Original Image')
subplot(2,2,2), imhist(I2(:,:,1))
    title('Adjusted Image, Gamma=1.0')
subplot(2,2,3), imhist(I3(:,:,1))
    title('Adjusted Image, Gamma=0.5')
subplot(2,2,4), imhist(I4(:,:,1))
    title('Adjusted Image, Gamma=1.5')
```

The result obtained using `imadjust` differs from that obtained using `histeq` (which we used in ▶Sect. 7.3 to enhance the contrast in the image). The function `histeq(I1,n)` transforms the intensity of image `I1`, returning in `I5` an intensity image with `n` discrete levels. A roughly equal number of pixels is ascribed to each of the `n` levels in `I5`, so that the histogram of `I5` is approximately flat. Histogram equalization using `histeq` to be carried out separately for each color, since `histeq` only works for two-dimensional images. We use `n=50` in our exercise, which is slightly below the default value of `n=64`.

```
I5(:,:,1) = histeq(I1(:,:,1),50);
I5(:,:,2) = histeq(I1(:,:,2),50);
I5(:,:,3) = histeq(I1(:,:,3),50);

subplot(2,2,1), imshow(I1)
    title('Original Image')
subplot(2,2,3), imhist(I1(:,:,1))
    title('Original Image')
subplot(2,2,2), imshow(I5)
    title('Enhanced Image')
subplot(2,2,4), imhist(I5(:,:,1))
    title('Enhanced Image')
```

The resulting image looks quite disappointing and we therefore use the improved function `adapthisteq` instead of `histeq`. This function uses the contrast-limited adaptive histogram equalization (CLAHE) by Zuiderveld (1994). Unlike `histeq` and `imadjust`, the algorithm works on small regions (or tiles) of the image, rather than on the entire image. The neighboring tiles are then combined using bilinear interpolation to eliminate edge effects.

```
I6(:,:,1) = adapthisteq(I1(:,:,1));
I6(:,:,2) = adapthisteq(I1(:,:,2));
I6(:,:,3) = adapthisteq(I1(:,:,3));

subplot(2,2,1), imshow(I1)
    title('Original Image')
subplot(2,2,3), imhist(I1(:,:,1))
    title('Original Image')
subplot(2,2,2), imshow(I6)
    title('Enhanced Image')
subplot(2,2,4), imhist(I6(:,:,1))
    title('Enhanced Image')
```

The result looks slightly better than that obtained using `histeq`. However, all three functions for image enhancement, `imadjust`, `histeq` and `adapthisteq`, provide numerous ways to manipulate the final outcome. *The Image Processing Toolbox—User's Guide* (MathWorks 2016) and the excellent book by Gonzalez et al. (2009) provide more detailed introductions to the use of the various parameters available and the corresponding values of the image enhancement functions.

The Image Processing Toolbox also includes numerous functions for 2D filtering of images. Some of the methods we have looked at in ▶Chap. 6 when filtering digital terrain models. The most popular 2D filters for images are Gaussian filters and median filters, as well as filters for image sharpening. Both Gaussian and median filters are used to smooth an image, mostly with the aim of reducing the amount of noise. In most examples the signal-to-noise ratio is unknown and adaptive filters are therefore used for noise reduction. A Gaussian filter can be designed using

```
h = fspecial('gaussian',20,10);
I7 = imfilter(I1,h);
```

where `fspecial` creates predefined 2D filters, such as moving average, disk, or Gaussian filters. The Gaussian filter weights `h` are calculated using `fspecial('gaussian',20,10)`, where `20` corresponds the size of a 20-by-20 pixel filter following the shape of a Gaussian probability density function with a standard deviation of `10`. Next, we calculate `I8`, which is a median-filtered version of `I1`.

```
I8(:,:,1) = medfilt2(I1(:,:,1),[20 20]);
I8(:,:,2) = medfilt2(I1(:,:,2),[20 20]);
I8(:,:,3) = medfilt2(I1(:,:,3),[20 20]);
```

Since `medfilt2` only works for two-dimensional data, we again apply the filter separately to each color (red, green and blue). The filter output pixels are the medians of the 20-by-20 neighborhoods around the corresponding pixels in the input image.

The third filter example deals with sharpening an image using `imsharpen`.

```
I9 = imsharpen(I1);
```

This function calculates the Gaussian lowpass filtered version of the image, which is used as an unsharp mask, i.e., the sharpened version of the image is calculated by subtracting the blurred filtered version from the original image. The function comes with several parameters that control the ability of the Gaussian filter to blur the image and the strength of the sharpening effect, and a threshold specifying the minimum contrast required for a pixel to be considered an edge pixel and sharpened by unsharp masking. Comparing the results of the three filtering exercises with the original image

```
subplot(2,2,1), imshow(I1)
    title('Original Image')
subplot(2,2,2), imshow(I7)
    title('Gaussian Filter')
subplot(2,2,3), imshow(I8)
    title('Median Filter')
subplot(2,2,4), imshow(I9)
    title('Sharpening Filter')
```

clearly demonstrates the effect of the 2D filters. As an alternative to these time-domain filters, we can also design 2D filters with a specific frequency response. Again, the book by Gonzalez et al. (2009) provides an overview of 2D frequency-selective filtering for images, including functions used to generate such filters. The authors also demonstrate the use of a 2D Butterworth lowpass filter in image processing applications.

We next rectify the image, i.e., we correct the image distortion by transforming it to a rectangular coordinate system using a script that is similar to that used for georeferencing satellite images in ▶Sect. 7.4. This we achieve by defining four points within the image, which are actually at the corners of a rectangular area (which is our reference area). We first define the *upper left*, *lower left*, *upper right*, and *lower right* corners of the reference area, and then press return. Note that it is important to pick the coordinates of the corners in this particular order. In this instance we use the original image I1, but we could also use any other enhanced version of the image from the previous exercises. As an example we can click the left side of the ruler at 1.5 and at 4.5 cm, where two thin white sediment layers cross the ruler, for use as the upper-left and lower-left corners. We then choose the upper-right and lower-right corners, further to the right of the ruler but also lying on the same two white sediment layers,

```
imshow(I1)
inputpoints = ginput
```

and click return which yields

```
inputpoints =
   418.2500   506.5000
   412.2500   728.5000
   773.7500   517.0000
   775.2500   737.5000
```

or any similar values. The image and the reference points are then displayed in the same figure window.

```
close all
imshow(I1)
hold on
line(inputpoints(:,1),inputpoints(:,2),...
        'LineStyle','none',...
        'Marker','+',...
        'MarkerSize',48,...
        'Color','b')
hold off
```

We arbitrarily choose new coordinates for the four reference points, which are now on the corners of a rectangle. To preserve the aspect ratio of the image, we select numbers that are the means of the differences between the x- and y-values of the reference points in inputpoints.

```
dx = (inputpoints(3,1) + inputpoints(4,1))/2- ...
     (inputpoints(1,1) + inputpoints(2,1))/2
dy = (inputpoints(2,2) + inputpoints(4,2))/2- ...
     (inputpoints(1,2) + inputpoints(3,2))/2
```

which yields

```
dx =
   359.2500

dy =
   221.2500
```

We therefore choose

```
basepoints(1,:) = [0 0];
basepoints(2,:) = [0 dy];
basepoints(3,:) = [dx 0];
basepoints(4,:) = [dx dy];
```

The function `fitgeotrans` now takes the pairs of control points, `inputpoints` and `basepoints`, and uses them to infer a spatial transformation matrix `tform` using the transformation type `projective`.

```
tform = fitgeotrans(inputpoints,basepoints,'projective');
```

We next need to estimate the spatial limits for the output, `XBounds` and `YBounds`, corresponding to the projective transformation `tform`, and a set of spatial limits for the input `xLimitsIn` and `yLimitsIn`.

```
xLimitsIn = 0.5 + [0 size(I1,2)]
yLimitsIn = 0.5 + [0 size(I1,1)]

[XBounds,YBounds]  = outputLimits(tform,xLimitsIn,yLimitsIn)
```

Then we use `imref2d` to reference the image to world coordinates.

```
Rout = imref2d(size(I1),XBounds,YBounds)
```

An `imref2d` object `Rout` encapsulates the relationship between the intrinsic coordinates anchored to the rows and columns of the image and the spatial location of the same row and column locations in a world coordinate system. Finally, the projective transformation can be applied to the original RGB composite `I1` in order to obtain a rectified version of the image (`I10`).

```
I10 = imwarp(I1,tform,'OutputView',Rout);
```

We now compare the original image `I1` with the rectified version `I10`.

```
subplot(2,1,1), imshow(I1), title('Original Image')
subplot(2,1,2), imshow(I10), title('Rectified Image')
```

We see that the rectified image has black areas at the corners. We can remove these black areas by cropping the image using `imcrop`.

```
I11 = imcrop(I10);

subplot(2,1,1), imshow(I1), title('Original Image')
subplot(2,1,2), imshow(I11), title('Rectified and Cropped Image')
```

The function `imcrop` creates displays of the image with a resizable rectangular tool that can be interactively positioned and manipulated using the mouse. After manipulating the tool into the desired position, the image is cropped by either double clicking on the tool or choosing *Crop Image* from the tool's context menu. The result of our image enhancement experiment can now be used in the next section to analyze the colors of individual sediment layers.

Recommended Reading

Abrams M, Hook S (2002) ASTER user handbook—version 2. Jet Propulsion Laboratory and EROS Data Center, Sioux Falls

Gonzalez RC, Woods RE, Eddins SL (2009) Digital image processing using MATLAB, 2nd edn. Gatesmark Publishing, LLC

Mathworks (2016) Image Processing Toolbox—User's Guide. The MathWorks, Natick, MA

Trauth MH, Bookhagen B, Marwan N, Strecker MR (2003) Multiple landslide clusters record Quaternary climate changes in the NW Argentine Andes. Palaeogeogr Palaeoclimatol Palaeoecol 194:109–121

Trauth MH, Strecker MR (1999) Formation of landslide-dammed lakes during a wet period between 40,000–25,000 yr B.P. in northwestern Argentina. Palaeogeogr Palaeoclimatol Palaeoecol 153:277–287

Zuiderveld K (1994) Contrast limited adaptive histograph equalization. Academic Press Professional, Graphic Gems IV. San Diego, pp 474–485

Editing Graphics, Text, and Tables

Electronic Supplementary Material The online version of this chapter (https://doi.org/10.1007/978-3-662-56203-1_8) contains supplementary material, which is available to authorized users.

8.1 Introduction

In ►Chaps. 2 and 3 we extracted text and tables from journal articles, webpages, and online data bases. In ►Chaps. 5 and 6 we then created various simple line graphs, bar plots, and block diagrams with MATLAB. In ►Chap. 7 we processed images, in particular satellite imagery. Both the vector graphics created in ►Chap. 5 and the raster graphics generated in ►Chaps. 6 and 7 require further editing before they can be published in journal articles or books, presented as elements of posters, or included in conference presentations. In this chapter we demonstrate how to edit vector graphics, raster graphics, text, and tables using both open source and commercial software packages. ►Section 8.2 is on editing and improving simple line graphs, as an example of vector graphics. The properties of the MATLAB figures, such as their line and fill colors, background color, stroke width and appearance, font type and size, and other properties, are modified to improve the legibility of the graphics. ►Section 8.3 demonstrates how to improve raster images by changing colors, color mode, resolution and other properties, and how to export the results in a specific file format that will depend on the type of document in which the image is to be used. ►Section 8.4 is about formatting text, structuring and tagging manuscripts, consistent use of paragraph and character styles. ►Section 8.5 is about editing tabulated data to prepare it for further formatting.

The following section provides a very brief introduction to the use of typical graphics editors by means of step-by-step tutorials, supplemented by example files documenting the workflow that are available online through Springer Extras (►http:// extras.springer.com). In this chapter all files with filenames ending in ..._matlab.eps were created with MATLAB in ►Chaps. 5 or 6 and are therefore stored in the materials folders of these chapters. The software used in this chapter was introduced in ►Chap. 1 and includes state-of-the-art graphics tools such as *Adobe Illustrator* and *Photoshop*, as well as widely used and very popular *Microsoft Word* and *Apple Pages*. After learning how to edit vector and raster graphics, texts, and tables we will then be ready to move on to the last four chapters (►Chaps. 9–12), which are on presentations, posters, manuscripts and multimedia publications.

8.2 Editing Vector Graphics

For our first example in this chapter we will demonstrate how to edit vector graphics. Vector graphics use geometrical primitives such as polygons, patches (or filled polygons), lines, or points. Examples of vector graphics are the line graphs, bar plots, and pie charts created in ►Chap. 5, and the coastline plot from ►Chap. 6. In MATLAB, simple lists of *xy* coordinates for *NaN*-separated polygons (such as the *age-temperature* data from ►Sect. 5.2, or the shoreline data set introduced in ►Sect. 6.2) form the basis of vector graphics. When using MATLAB or any other software to visualize such data, the graphics functions display plots of the *xy* coordinates as complex vector graphs with multiple graphics properties that include line thicknesses and colors, markers, labels, background colors, coordinate axes, and so forth. Vector graphics software then allows the properties to be altered and other, more sophisticated properties to be added to be graphics.

8.2.1 Line Graphs: Getting Started, Panels, Layers, and Tools

In this subsection we will first edit the line graph from ▶Sect. 5.2 showing temperature and snow accumulation variations in the GISP2 data from Alley (2000). Comparing the original line plot created using MATLAB with the totally redesigned versions shown in ◻Fig. 8.1 gives an idea of the possibilities described in this section.

We use the commercial *Adobe Illustrator* to demonstrate some of the basics of editing vector graphics. All vector graphics in this book have been created with the English edition of Adobe Creative Cloud. Our tutorial, however, should also work in a very similar way with earlier versions of the software, or with versions in other languages, or other software. Further information about the software and technical terms or jargon in italics can be found at ▶https://helpx.adobe.com/illustrator/topics.html. Adobe Illustrator allows us to import the EPS file *icecore_lineplot_vs1_matlab.eps* created with MATLAB in ▶Chap. 5 using *Open File* from *File* menu. Most vector graphics

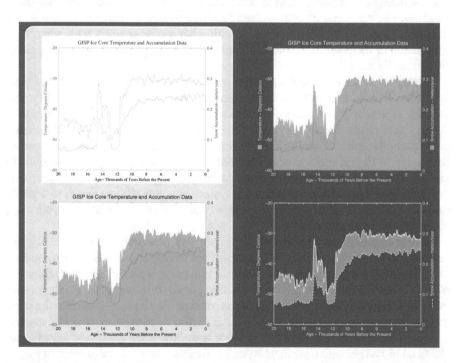

◻ **Fig. 8.1** Variations in the GISP2 data from Alley (2000). The original MATLAB figure *icecore_lineplot_vs1_matlab.eps* is presented in the upper left hand column of a dummy poster. It shows the line drawing with title, axis labels, axes, and graph on a white background, which is the standard template. In the lower left corner of the poster the filled graph is shown with white background removed. In the upper right corner the graph is shown with white instead of black for the axes and labels, to be used on a dark background for a poster or presentation. In the lower right corner a decorative version of the graph is shown for use in leaflets, with white labels and axes instead of black

software tools should not have a problem with opening an EPS file; in the rare case of incompatibility problems we could also create PS or PDF files with MATLAB and import these, again using *Open File*.

The line graph is displayed in a new figure window. At the top of this window we find *Application bar*. In this drop-down menu we find a number of push-buttons assembled in a toolbar. These provide access to all kinds of information about the document and also include various shortcut buttons for some frequently-used functions and various context commands. On the left side of the window we find *Tools panel*. To become familiar with the names and functions in *Tools panel* we can move the mouse over their buttons to make the tool tips visible and we can also obtain some information on the functions by browsing *Adobe Support > Illustrator > Tools*. Having opened the *icecore_lineplot_vs1_matlab.eps* file, we immediately save the document under a new name using *Save As* from the *File* menu. We call our Illustrator file *icecore_lineplot_vs2_ai.eps*. It is recommended that the original MATLAB file should always be retained as we may wish to return to it at a later stage to create another version of the graph. Both Inkscape and Illustrator allow us to save the file in various vector file formats, such as EPS, PS or PDF, and these are the formats that we will use because they are compatible with most other vector graphics programs, whereas vector file types that are native to particular types of software can often not be imported into other programs. Illustrator has its own native file format, which is identified by the *.ai* file extension. Illustrator files are compatible with other vector graphics tools when PDF information of the graph is included in the file; this is an optional feature when creating files in Illustrator.

During all stages of a graphics design project it is always important to save your work from time to time using *Save as,* since any graphics software may crash unexpectedly, especially when using older versions of the Microsoft Windows operating system. More recent operating systems provide an automatic save every few minutes, making it unnecessary to save your files manually. Backing up your work onto separate disks every 5–10 min is however still advisable, using, for example, *Time Machine* in macOS, or another automated backup system. It is also a good idea to always keep earlier versions of a graph while working on it, since you may wish to return to a previous layout. Again, modern operating systems create different versions of all your files, which you can recall if necessary.

Having imported and saved the graph, we will now start editing the file. We first need to set up our working environment (or workspace) so that we can quickly access the most frequently used functions. In Illustrator, we select *Essentials* from the *Window > Workspace* menu to open all the panels that we wish to work with (◻Fig. 8.2). Taking a closer look at the *Layers* panel in Illustrator, we note that the three original graphics layers created by MATLAB (*figure window, axes,* and *plot objects*) were unfortunately not preserved as separate layers when the graph was saved as an EPS file, but merged into one single layer. For further editing of the graphics we need to move all different types of object into individual layers. The advantage of working with layers is that the different types of objects can be selected and edited individually by toggling the lock and visibility symbols of each layer. Illustrator also displays a list of all the objects within a single layer in the *Layers* panel when you click on the gray > next to the layer's name, e.g., next to *Layer 1*. As you can see, there are many

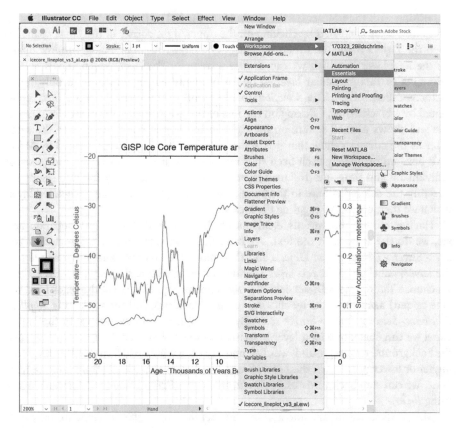

◻ Fig. 8.2 Screenshot of the icecore line graph, showing Adobe Illustrator while choosing *Workspace > Essentials*. For further informations browse ▶https://helpx.adobe.com for *Workspace basics*

different types of graphics elements in *Layer 1*, such as <Text>, <Path> and <Group>; these can be selected individually by clicking the corresponding element in the list (e.g. <Path>), which then becomes active and is marked with a small square.

Editing a graph generally includes the following steps: (1) creating and naming layers for different graphics objects, such as the text, the line for snow accumulation, the line for temperature, the coordinate system (with ticks and axis labels), the background, and the frame; (2) selecting and dragging the individual objects into the specific layers; (3) editing the white background and frame; (4) changing the properties of the graphics objects, such as their line styles, fillings, and colors; (5) editing text objects in the layers, e.g., choosing serif or sans-serif fonts and font sizes; (6) adjusting technical properties affecting the printing process, such as the handling of transparency and the color black; and finally (7) saving the resulting graph as a PDF or an EPS file.

The *Layers* panel of Illustrator allows graphics objects belonging to individual layers to be organized and viewed in a series of sublayers. These sublayers can be viewed by clicking with the pointer on the symbol > to the left of the particular layer of inter-est. A new layer is created by clicking on the upper right corner of the *Layers* panel to open the *Layers* menu, and then selecting *New Layer*. A small dialog window opens for the new layer in which we can choose a color for the layer, which allows objects belonging to this particular layer to be easily identified. In Illustrator, a polygon or vector is called a *Path*. The *Selection Tool* is found in the vertical *Tools Panel*, repre-sented by a black arrow. Clicking on the line of the graph makes the *curve anchors* vis-ible and shows a small square on the right in the *Layers* panel, filled with the same color as the curve to show that it is active. We find that the curves created by MAT-LAB have unfortunately each been split into three segments, which we need to select individually and drag into the new layer. We do this by selecting the line segments, selecting the colored square in the panel, and moving both into the designated layer, which we then protect by toggling the lock symbol.

Having moved the various elements of the axes and curves into separate lay-ers, we do the same with the text. We can select text very quickly using *Application bar > Select > Object > Text Objects* and then move the colored square that appears in the *Layers* panel up to the *text* layer. Having first ensured that the *graph* layer is locked, we choose *Tools panel > Direct Selection Tool* and very quickly select the black (and the blue or red) axes and ticks by clicking on one black tick and then selecting *Applica-tion bar > Select > Same > Stroke color*. By toggling the visibility symbol in the *Layers* panel we can check in which layer a particular object is located. To move one layer to the background or foreground, we click and drag the whole panel row of this layer to a higher or lower position in the *Layers* panel. If we want to undo anything that we have done, we can step backwards using *Application bar > Edit > Undo* (and forwards again using *Redo*).

MATLAB seems to take into account the fact that the *PostScript* standard only sup-ports certain specific techniques for implementing transparency. In Illustrator, the EPS document contains two white rectangles. We move the larger of these two rectangles to a layer named *frame*, which we then move down to the bottom of the list of lay-ers. The smaller rectangle has the same dimensions as the coordinate axes. We move it to the *back* layer, which we then move to the second lowest position in the list of layers. We can also edit the appearance of the two rectangles in *back* and *frame* by, for instance, changing the *Fill* color or the *Stroke* color, or modifying the sizes of the rectangles. After moving the two rectangles into the *back* and *frame layers*, we can lock these layers and then hide them by switching off their visibility. We save the line draw-ing once again as *icecore_lineplot_vs2_ai.eps*.

We next continue with editing the curves themselves by, for example, filling in the area between the curves and the *x*-axis of the graph (▪ Fig. 8.1). Before we are able to fill in this area we might need to join together the segments of the curves. Splitting into separate segments is sometimes an unfortunate characteristic of polygons cre-ated by the *PostScript* export plugins of many types of software, including MATLAB. Merging polygon segments is performed in Illustrator by selecting the segments of the blue line and then combining them using *Application bar > Object > Compound Path > Make*. Alternatively, to merge two segments of the blue polygon, we can use the *Direct Selection* tool and select one of the segments. Using the *Lasso* tool we carefully select the terminal vertex at the end of the segment of the blue line and the vertex at the beginning of the adjacent segment. Using *Application bar* > *Object* > *Path* > *Join* we combine both segments into a single segment and then repeat the process with the remaining third segment. The segments of the red curve are then linked using the same procedure. As the final step in editing the line graph we will now fill in the area beneath the curves. For this we duplicate the *graph* layer using *Layer Panel Menu* > *Duplicate*, rename the new layer *graph fill*, put the layer into the third-lowest position in the list of layers, unlock the new layer, and lock all other layers.

To create a closed polygon we now give a demonstration of how to use Illustrator to create new polygons, which of course is the classic application of vector graphics software tools. We choose *Pen Tool* from the *Tools panel* and select the left terminal vertex of the blue curve, which is indicated by/*anchor* when hovering over this vertex. Holding down the *Shift* key on the keyboard allows the creation of a new vertex located either vertically at a 90° angle above or below the terminal vertex, or horizontally at a 180° angle to the left or right of the terminal vertex. We decide to place the new vertex at the lower end of the red axis, on the left hand side. While still holding down the *Shift* key, we click on the right end of the black *x*-axis and a third time at the same height as the right end of the blue curve, trying to meet its terminal vertex. Taking a closer look at the snow accumulation curve we note that there is no data point for zero age, i.e. a small gap exists between the blue curve and the blue *y*-axis to the right, which means that the vertical line does not meet the right end of the blue curve. To close this gap we connect the blue curve and our new vertical line using the *Lasso* tool, as described above.

Next, to select a color shade for the *Fill* that goes together nicely with the blue of the curve, we use the *Swatches* panel and the *Color Guide* panel. The *Swatches* panel should already be visible in our workspace. We can browse all shades of colors using *Application bar* > *Window* > *Color Guide*. We again select the blue curve and then the *Tools* panel shows which color is used for the *Stroke* of our object. By clicking on the small arrow in the upper right corner (next to the symbols for stroke and fill in the *Tools* panel) we switch between stroke and fill to apply an intense blue color to the fill. As the color is a bit too intense, we select a paler shade from the *Color Guide* panel and the fill, still being active, changes color.

We then transfer the colors in our document to the *Swatches* panel. To do this we unlock all layers using the *Layers* panel menu and *Application bar* > *Select* > *All*, after which the colored symbols for each layer appear as small squares. We click on the upper right corner of the *Swatches* panel, open its *Context* menu, and choose *Add Used Colors,* which automatically adds all colors used in the document, each represented by a square color field. We continue by creating the color fill for the area beneath the red curve and choosing a suitable shade, as we did for the blue curve. We save the file as *icecore_lineplot_vs3_ai_fill.eps*.

If the graph is to be published in a research article the printing costs can be reduced by using a black and white (b/w) version. Printed versions of papers today often contain b/w illustrations, with colored versions included in the PDF version of the paper published online. It is important to note that a figure in *RGB* color mode, which is used in digital versions to be read on a computer screen need to be converted into the *CMYK* color mode if the figure is being provided for offset printing. In other words, the black color should be a pure black (also known as *key*) and not a composite of maximum intensity red, green, and blue in the RGB mode. We can change the color mode into CMYK using *Application bar > Edit > Color Mode* and then check the two black color swatch buttons in the *Swatches* panel. Double clicking on a color swatch opens a panel called *Swatches Options.* We find that the ticks and *x*-axes use the correct color K = 100 (for 100%) and remove the other black color by selecting it and then clicking on the trash icon in the lower right corner of the panel. We then select any of the black text objects and automatically select all others using *Application bar > Select > Same > Fill Color*, and then use K = 100 for all text in the graph. It might even be appropriate to choose *Color Mode > Grayscale* in the *Swatch Options* window. If we do not change RGB black into CMYK black (or *key*), all black text will appear as a smeared mix of the three colors, red, green and blue, since the ability to print the three colors precisely on top of each other is very rare in professional printing. We adjust the color values for red and blue to whole numbers and save the file as *icecore_lineplot_vs4_ai_cmyk.eps.*

Continuing with the grayscale version of the graph *icecore_lineplot_vs3_ai_fill. eps*, we select all graphics objects, such as coordinate axes and ticks, and change their stroke color into CMYK black (or *key*). Double clicking on the color swatch in the *Swatches* panel opens a small window called *Swatches Options.* Instead of *Color Type > Process Color* we now choose *Grayscale.* To deactivate all objects after we have completed this task, we click with the pointer on an empty space in the work area. Since there is now no color information available to distinguish between the two curves, which were originally red and blue, we select the *Snow Accumulation* curve and change the line to a dashed line using the *Stroke Style* dialog box of the *Stroke* panel. Alternatively, we could again use different shades of gray for the filled polygons, i.e. the areas below the curves, to distinguish between the two curves. In order to associate each of the two curves with its corresponding *y*-axis, we can either label the curves or add a legend. We then save the file as *icecore_lineplot_vs5_ai_grayscale.eps.*

For use with a dark background we change all text and black contours to white, as shown in *icecore_lineplot_vs6_ai_fordarkbackground.tif.* In this case we open the *Attributes panel* by *Application bar > Window > Attributes* to make sure that the checkboxes for fill and stroke (*Overprint Fill* and *Overprint Stroke*) are disabled. More information about *overprinting* are given in the next section.

We have demonstrated a typical workflow for editing a line graph, which can also be used to edit other *xy* and *xyy* plots. As an example, the bar plot created in ▶Sect. 5.3 and stored in *icecore_bargraph_vs1_matlab.eps* essentially comprises similar graphics objects, such as the axes, text objects, and the white background, but with filled rectangles with black outlines.

8.2.2 Pie Charts: Transparency, Raster with Vector, and Advanced Features

The pie chart from ▶Chap. 5 in *icecore_piechart_vs1_matlab.eps* visualizing 2D data in 3D looks very attractive when using a transparency effect. However, if we have already generated transparency in MATLAB the program unfortunately exports a raster file (even though it was saved as the EPS file *icecore_piechart_vs2_matlab.eps*) and the vector information is lost. Of course for many applications a raster (or bitmap) file is easier to handle in a production process than a vector file. However, disadvantages of the raster format include the greater file size, the fact that no further editing of text or other objects is possible, the fixed resolution, and the very limited cropping and editing options. We might therefore prefer to export an EPS file from MATLAB with opaque colors, such as in *icecore_piechart_vs1_matlab.eps*, and then add transparency with the vector graphics software tool (◘Fig. 8.3). To demonstrate this we open the file in Illustrator, choose *Tools panel > Direct Selection Tool*, select all objects using *Application bar > Select > All*, and immediately reveal the complex structure of this chart. With all objects still selected we click on the *Swatches* panel, open its context menu, and choose *Add Selected Colors*. To deactivate all selections we click with the pointer over empty space of the work area.

Using *Tools panel > Direct Selection Tool* we click and drag the text object *>1%* a few millimeters to the right to avoid overlapping text. If we aim to publish the

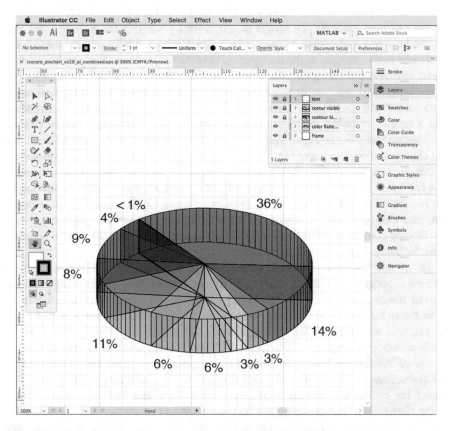

◘ **Fig. 8.3** Screenshot of the icecore pie chart, showing the Adobe Illustrator workspace with *Tabbed Document window, Application bar, Workspace switcher, Panel title bar, Control panel, Tools panel,* the *Layers* panel, and the vertical *Dock* with 14 icons representing collapsed panels organized in panel groups

graphic in a research article, we can change the color mode for the whole document into CMYK using *Application bar > Document Color Mode > CMYK*. We then select all text objects and move them into a separate layer called *text*, using *Application bar > Object > Text Objects*. With all text objects still selected, we change the color mode in the *Color* panel to CMYK and find that the black color for text objects is still RGB black, consisting of a mix of the three colors red, green and blue, as in the previous graph. We therefore again have to change the color composition by moving the scrollbar of each color channel to the values $C = 0$ $M = 0$ $Y = 0$ $K = 100$, or alternatively, by typing the values for CMYK manually into the corresponding text fields. Clicking on the color field in the *Swatches* panel opens a small window entitled *Swatch Options*. Instead of *Color Mode > CMYK* we now choose *Grayscale*, as we did in the previous example. To deactivate all objects we click the pointer over an empty space in the work area. We then select all black outlines using the *Direct Selection Tool* and use *Application bar > Select > Same > Stroke color*, after which we move the outlines into a new layer called *contours*. The black outline color is again RGB black and we therefore have to delete the mixed RGB color in the *Swatches* panel. All contours are changed into the correct grayscale black by clicking on the correct black color field.

We can now examine the results of our work by toggling the visibility and lock for the new *text* and *contours* layers. Using *Application bar > Select > All* the colored zones and two rectangles in the background are selected, as in the line graph discussed previously. We again drag the rectangles into a new layer called *back*, move this layer to the lowermost position in the *Layers* panel, and lock the layer. Using *Application bar > Select > All* should now only select the colored zones and we then move them into a new layer called *color*. The original layer is now empty and can be deleted. Clicking on the *color* layer we again activate all colored objects in our pie chart and can change the opacity using the *Transparency* panel, either by typing *50% Opacity* or by dragging the scrollbar to the appropriate value. We then save the file using *Application bar > File > Save As*, choose *Medium Resolution* or *High Resolution* depending on our requirements, activate the *Compatible Gradient and Gradient Mesh Printing* checkbox, and then save the file under the name of *icecore_piechart_vs4_ai_transparency. eps*. Having completed all of these steps the pie chart is now ready for digital printing, or for inclusion in any type of presentation.

Let us now continue with some of the advanced features offered by Illustrator for creating files suitable for offset printing. In theory, creating files by flattening gradients such as transparency is always possible by exporting the whole graphic as a bitmap, or in raster form with the required resolution, using *Application bar > File > Export*, for example as the JPEG file *icecore_piechart_vs5_ai.jpg*. A second possible way to create a vector file (which is not recommended in every case) is to transform each area of different color into an individual object using *Application bar > Object > Flatten Transparency*. The third and safest method to use for offset printing is to first export the critical transparent areas into a bitmap file and then later replace them in the illustration, thus retaining the text and contours as vectors. We continue with the *icecore_piechart_vs4_ai_transparency.eps* file that we created previously and toggle the visibility so that only the *color* layer remains visible. Using *Application bar > File > Export...* we export the file as, for instance, a Portable Network Graphics (PNG) file with a high resolution. We then toggle the visibility again so that only the *color* layer remains hidden. This layer will be eliminated at a later stage. We then form a new layer called *color bitmap* and import our PNG file into it using *Application bar > File > Place...*, move the image to the correct position so that it replaces the image in the *color* layer, and save the file as *icecore_piechart_vs10_ai_combined.eps*.

As a second example of an advanced application let us create a cropped image from our pie chart using the *clipping path* feature in Illustrator. We first need to define the outline of the cropped image using the *Rectangle tool* from the *Tools panel*. With this tool we draw a rectangular outline around the diagram, click on one of its corners, and then drag the cursor diagonally until the rectangle has the same width as the pie chart but a slightly lower height. Having completed the rectangle, we use the *Add Anchor* tool from the *Tools panel* to add an anchor point to the middle of each of the two horizontal sides of the rectangle and then drag these anchor points to the upper and lower edges of the pie chart. To convert this irregular-shaped hexagon into the pie shape we then use the *Convert Anchor Point* tool from the *Tools panel*, click on the upper point, and drag the handle of the *Bezier* curve to transform the horizontal line into a semicircle. After also adjusting the lower horizontal line, we select this irregular-shaped hexagon and the PNG image of the pie. Using *Application bar > Object > Clipping Mask > Make* then creates a *clipping path* following the outline of the PNG image, which cuts away the white background that surrounds the original image. To see whether the *clipping path* does in fact accurately cut away the white parts we can draw a random shape with an arbitrary fill in the background and delete it after this check. Finally, we save our file in an EPS format as *icecore_piechart_vs11_ai_mask.eps* (◾Fig. 8.3).

Our third advanced example demonstrates *overprinting* of black contours and text objects during the offset printing process. Using *Application bar > Window > Separations Preview* we open a panel displaying the four printing plates for *Cyan, Magenta, Yellow and Key/Black* (CMYK). If we hide the black printing plate we notice that, in little gaps between the colored areas, lines and text that were previously black now appear white in the *Separations Preview*. Since the four printing plates can become slightly displaced relative to each other during printing, the final product can have black lines or text with undesirable white gaps. To avoid this problem we select all the black objects in our illustration by first selecting any arbitrary black object and then using *Application bar > Select > Same > Stroke Color* to select all others. Using *Application bar > Window > Attributes* we then open a new panel in which we click on the *Overprint Stroke* checkbox. The result can be seen by again hiding the black printing plate in the *Separations Preview* panel: the colored areas are now continuous beneath the black contours. Similarly, we add the overprinting feature to the black text objects by activating *Overprint Fill* checkbox in the *Attributes* panel and save the final product as a file with the name *icecore_piechart_vs12_ai_overprint.eps*. Please note that the overprinting attribute needs to be handled with care. In most cases it is only applied to black objects since overprinting for lighter colors, in particular for white, leads to the complete invisibility of objects in these colors.

As our last example of the advanced features in Illustrator we save the colors used in this particular file for use in future applications, using the *Adobe Swatch Exchange* (ASE) format. We first open the *Color* panel menu by clicking on the *Menu Icon*, which is the second symbol in the upper right corner of the *Color* panel. Choosing

Add Used Colors transfers all swatches into our panel. Opening the *Color* panel menu again, we choose *Small List View* and now find all colors listed by their names, for instance $C = 0\ M = 0\ Y = 0\ K = 100$ for the black color, named after the color separation. Clicking on the black color in the *Color* panel opens the *Swatch Options* window to allow various options to be checked or adjusted; for example once again choose grayscale for the $K = 100$ color. Back to the *Color* panel menu, we click on *Save Swatch Library as ASE* to save the colors as *icecore_piechart_swatches.ase* so that this *Swatch Library* can be transferred to other graphics files. If we want to load a *Swatch Library* into another graphics file, we select *Open Swatch Library* in the menu and search for the desired file on our computer or use the built-in library. The swatch library saved in *icecore_piechart_swatches.ase* is now available for use with other applications within Adobe Creative Cloud such as Illustrator, InDesign, and Photoshop, or can be handed over to another colleague for further processing.

8.2.3 Rose Diagrams: Manipulating Clipping Paths

The rose diagram in the *directional_rosediagram_vs1_matlab.eps* file reveals that the MATLAB export routine has created a rather complex assemblage of graphics objects such as lines, patches, and background rectangles, which will have to be edited before they can be included in a publication or presentation. Clicking on the triangle symbols on the left side of the *Layer* panel shows one group that includes a red polygon, text elements, and four other groups containing various graphics objects, various paths, and the clipping paths. We select and organize the text and the red polygon into separate new layers called *text* and *graph*, as we did in the previous sections.

Zooming into the upper and lower parts of the diagram we find that the outer circle extends beyond one of the object's frames, or more precisely, beyond the limits of one of the clipping paths. By toggling the visibility of each of the layers one by one, we check in which layers the different objects are located and find that hiding all clipping paths will not change the overall appearance but at least fixes the truncated outer circle. We create new layers called *dotted radiation*, *outline*, and *dotted circles* and move the relevant objects into these layers. All superfluous objects and clipping paths of the graphic remain in the original *Layer 1*, at the lowest level in the *Layer* panel. We toggle *Layer 1* to make it invisible; later it will be deleted.

Clicking on the *graph* layer activates the red polygon and we choose a white color for the fill. Opening the *Stroke* panel and clicking on the content of the *dotted circle* and *dotted radiation* layers we find that a *stroke weight* of 1 pt (points) has been used in combination with a *dashed line* that has a 0.5 pt dash and 4 pt gap, both of which settings we can change to other values in this panel. We check the appearance of the diagram, which overlies a colored background, by creating a new layer at the bottom of the list of layers and drawing a filled rectangle using *Tools panel > Rectangle Tool*. We then delete this layer once again. We might prefer a white background for the rose diagram circle: we therefore duplicate the *outline* layer and name it *white circle*. We then lock all layers except the new one, move it below the original *outline* layer, set the *white circle* contour to *None (/)*, and give the filling a white color. To complete our procedure we remove the *back* and *Layer 1* layers and save our file under the name *directional_* *rosediagram_vs1_ai_lines.eps*. For other purposes we can give the rose diagram circle a fill similar to that in the file named *directional_rosediagram_vs2_ai_fill1.eps* in the supplementary material accompanying this book, or define different colors for each segment of the rose diagram, as in the file called *directional_rosediagram_vs3_ai_fill2.eps*.

8.2.4 Shoreline Data Set: Number of Vertices in Vector Graphics

For our next example we will work on the shoreline plot created in ▶Sect. 6.2. We have already discussed the problem of large numbers of vertices (in ▶Sect. 6.2) and used a MATLAB script to halve the number of data points plotted. At first sight the graphics object in the *coastline_linegraph_vs1_matlab_28033vertices.eps* file looks very familiar after having worked on the line graph in *icecore_lineplot_vs1_matlab.eps*. The difference is that the shoreline data set has 28,033 longitudes and latitudes pairs, resulting in polygons with a total of 28,033 vertices in the graph. Despite the relatively small file size of 139 kB, this enormous number of vertices may cause problems when trying to print the polygons with a PostScript printer, and also when including the graph in slides or some other medium. For example, a PDF document containing such complex vector objects is difficult to browse through or print, and also difficult to handle as a source for offset printing. As an example we will consider Fig. 2 from a publication in Quaternary Science Reviews (Trauth et al. 2009),

```
Trauth, M.H., Larrasoaña, J.C., Mudelsee, M. (2009): Trends, rhythms
and events in Plio-Pleistocene African climate. Quaternary
Science Reviews, 28, 399-411.
```

which can be downloaded from the Science Direct webpage if you have electronic access to that journal:

```
http://sciencedirect.com/science/article/pii/S0277379108003302
```

There are three panels in this figure: A, B and C. Panels B and C comprise three curves for different paired window sizes, 15, 30 and 50 kyr. For mathematical reasons each of these curves consists of 5,000 data points, i.e. the polygons have 5,000 vertices. Having downloaded the file, the very tedious scrolling forward and backward across Figure 2 clearly demonstrates the problems of dealing with large numbers of vertices. Furthermore, printing the file is a great challenge, as well as being a time-consuming task on most printers. In this example the copy editing office of the publisher should have come back to the author to improve the graph for publication. In order to get around this problem a second version of the file has been created that includes a JPEG version of Figure 2 for easier scrolling and printing. Another example is ◻Fig. 5.20 in *MATLAB Recipes for Earth Sciences—3rd Edition* (Trauth 2010), which was originally delivered to the designer as a vector file with 5,000 vertices for each curve but subsequently converted into a TIFF file for publication.

To prepare a job for printing, most printing devices use a built-in *Raster Image Processor* or *RIP* software module to interpret, render, and screen vector or bitmap data, which essentially means to transform it into a high-resolution raster image. This process is called *ripping* or *rendering* vector data. If the processing of a complex vector file approaches or exceeds the limitations of the device then the printer may require an extremely long processing time, or even stop printing and return a *limit check* or *VMerror* (which stands for *Virtual Memory Error*) message. Editing such a complex vector file, or placing it in a desktop publishing document, or embedding it into PDF may also, in some cases, result in a serious loss of performance, depending on the available hardware and the quality of the software product used.

In most cases (e.g., when using the graph in a conference presentation, on a poster, or in a manuscript), processing of the original, high-resolution shoreline data set is not advisable. If you are the creator of the high-resolution figure in *coastline_linegraph_vs1_matlab_28033vertices.eps* it would be better to run the script from ▶Sect. 6.2, in order to reduce the number of vertices. For comparison, we first run the script and reduce the number of vertices by a factor of two, creating a plot with 14,017 vertices, which we save as *coastline_linegraph_vs2_matlab_14017vertices.eps*. A second run again halves the number of vertices to just 7,009, and we save this plot in a file called *coastline_linegraph_vs3_matlab_7009vertices.eps*. The original illustration can also be exported from MATLAB as a bitmap file of 600 dpi or, in our example, 300 dpi resolution, and saved as a file called *coastline_linegraph_vs1_matlab_28033vertices.tiff*.

Should we not have access to the original data set and therefore not be able to run the MATLAB script to reduce the number of vertices, we can instead use graphics software to reduce the number of vertices (referred to as *nodes* and anchor points in graphics software jargon). The *Adobe Support* website

```
https://helpx.adobe.com/illustrator/kb/create-efficient-illustrator-files-
improve.html
```

has many useful hints on how to create more efficient vector files, as well as information on the influence that specific graphics effects and elements have on memory requirements, and on potential problems when editing vector files. The performance of graphics software when editing a large vector file can, for example, be significantly improved by splitting complex outlines with many anchor points or many curves into shorter segments.

Let us now open *coastline_linegraph_vs1_matlab_28033vertices.eps* (with its 28,033 vertices) in Illustrator in order to optimize file performance by reducing the number of vertices in our graphics software. We choose *Object > Path > Simplify* to reduce the number of vertices. With the *Show Original* and *Preview* check boxes activated, we try different settings for *Curve Precision* and *Angle Threshold*. We can also try out the *Straight Lines* option to see if it is suitable. During this process we can observe the number of vertices in the *Original pts* and *Current pts* lines as well as the shape of the preview curve compared to the original.

According to the *Simplify* option, the original number of vertices in the file is 14,398, not the intended 28,033. When we worked on the 1st edition of this book more than five years ago, exporting EPS files from MATLAB created files with the correct number of vertices. Polygons at that time were split into smaller segments, which could be connected with each other using Illustrator. However, the most recent MATLAB release does not appear to split the polygons but instead reduces the number of vertices, as can be seen in the *coastline_linegraph_vs1_matlab_28033vertices.eps* example. When MATLAB support was asked about this the response was "We do some optimization when drawing line vertices, and reduction in vertices count in generated EPS file is result of that." It is important that readers should be aware that MAT-LAB reduces the number of vertices when saving EPS files.

8.2.5 Pseudocolor Plots: Combining Raster with Vector

The pseudocolor plot created in ►Sect. 6.3 depicts the ETOPO1 digital topography and bathymetry on a regular 1-min grid. Opening the file *etopo1_pseudocolorplot_vs1_matlab.eps* reveals that MATLAB generated a raster file instead of a vector file, and in so doing lost the vector information for the axes, ticks and labels. There are several ways around the problem of rasterizing a vector graph, including a pseudocolor layer. The first alternative is to create a raster version of the graph with disabled axes, ticks, and labels using the *axis off* command in MATLAB, to save this as a first version of the graph, and then to create an empty set of axes with only the axes, ticks, and labels and save this as a second version. Both files can then be combined using vector graphics software such as those described above, rather than having to redraw or retype the vector objects of the graph.

A second alternative is to create a new layer above the layer containing the original pseudocolor plot and then redraw and retype the axes in the graphics software. We can also crop the image by creating a *clipping path* in our vector software. Alternatively, we can crop the colored interior of the graph using pixel software such as *Adobe Photoshop*, save it as a JPEG or TIFF file, place this file into the vector file and save the illustration as *etopo1_pseudocolorplot_vs3_ai.eps*.

8.2.6 Block Diagrams: Three-Dimensional Visualization and Labeling

The ETOPO1 data set created in ►Sect. 6.3 and stored in the *etopo1_surfaceplotlight_vs1_matlab.jpg* file shows an attractive, detailed, and fully colored three-dimensional digital elevation model. We will refine this graphic by adding labels and presenting it as a three-dimensional block diagram (◻Fig. 8.4).

We first open the original image in Photoshop and reduce the document size to a width of 2,700 pixels only, which means that a 20 cm wide printout has a resolution of 300 dpi. We then cut the white background and save the graphic as *etopo1_surfaceplotlight_vs2_ps_cropped.jpg*. We now create a new page 100 mm^2 using *Adobe Illustrator* and emplace the cropped image using *Application Bar > File > Place... .* If the cropped image has a white frame around the actual graphic and is therefore slightly larger than 100 mm^2 we can carefully adjust the dimensions of the image by holding the *Shift* key down and the image will be automatically adjusted to fit the page. We save the preliminary version of the graphic in a file that we call *etopo1_surfaceplotlight_vs3_ai_raw.eps*.

We next create some labels to indicate important locations on the map. To do this we first create a new layer called *text*, choose *Tools panel > Type Tool* and type *East African Plateau*. We mark the text object using the *Character* panel, define *Helvetica Regular 8 pt* as the font type and size, and then drag the text frame into the correct position. To save time we duplicate this text object several times using *Copy &*

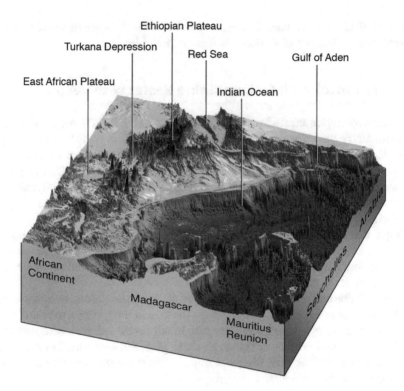

◘ Fig. 8.4   Block diagram with three-dimensional visualization and labeling

Paste, or alternatively, by dragging the text frame to another position while keeping the *Alt* key depressed. We then change the text of the various text objects to *Turkana Depression*, *Red Sea*, *Indian Ocean* and *Gulf of Aden*. Below the 3D surface we rotate the *African Continent*, *Madagascar* and *Mauritius Reunion* labels using *Application bar > Object > Transform > Rotate* and choose an *Angle* of −11°. We likewise type and rotate the *Seychelles* and *Arabia* labels, choose an *Angle* of 59°, and move all of the labels to their appropriate positions.

We next create a new layer called *box* for the three-dimensional block underneath the surface plot. We then select *Tools panel > Line Segment Tool*, press the *Shift* key, and draw a vertical black line 21 mm high with a *Stroke Weight* of 0.6 pt, which we move to the left edge of the graph. We make another two copies of this object, change the height to 8.6 mm for the edge on the right and to 13.7 mm for the edge in the foreground. We then create a horizontal line 73.6 mm long, rotate it to −11° (as for the labels above) and connect the bottom of the vertical line on the left with the bottom of the line in the center. A line 54 mm long and rotated by 59° connects the bottom of the vertical line in the center with the bottom of the line on the right.

Unfortunately, the lines defining the 3-D block are not yet visible since the *box* layer underlies the opaque *bitmap* layer containing the colored JPEG image that we placed there previously. In the next step of our editing process we will eliminate the white background. We find the file name *etopo1_surfaceplotlight_vs2_ps_cropped.jpg* written in blue on the left side of the *Control panel*. Clicking on this file name opens a *Context* menu in which we choose *Edit Original* to open the original file in our preferred image processing software, which in our case is Adobe *Photoshop*. Using *Application bar > Select > Color Range* in Photoshop we select the white color and set the *Fuzziness* to 100, which affects the falloff beyond the selected boundaries. In the *Layers* panel we click on the only layer present, which is called *Background,* and rename it as *Layer 0*. Using the *Backspace* key deletes the white zone and in its place we get a transparent zone around the image instead of the former opaque white background. Since JPEG files do not allow the use of transparency, we save the file as a PNG file called *etopo1_surfaceplotlight_vs5_ps.png* and return to Illustrator.

Using *Control panel > Linked File* and then clicking on the name of our original JPEG image and the small *Relink* icon at the bottom left, we establish a new link. While doing so we replace the previously emplaced JPEG image with our new PNG file. Using *Control panel > Embed* we then embed the PNG file. In the *Layers* panel we create a new layer called *strokes*, which we move together with the *text* layer to a position above *Layer 1* to ensure that they always remain visible. In the *strokes* layer we then connect the upper labels with the corresponding zones of the illustration by drawing vertical lines of 0.6 pt thickness. We then save the file as *etopo1_surfaceplotlight_vs6_ai_labels.eps.*

Let us again explore some advanced applications of the graphics software, starting with the use of color gradations. We first create a new layer called *box color zone* at the bottom of the list of layers for the color of the vertical surfaces of the block diagram. We then choose *Tools panel > Pen Tool*, select an arbitrary sandy, tan or gray color from the *Swatches* panel for the fill, and draw the south and east facing vertical surfaces of the block diagram. Since the *box color zone* layer is below *Layer 1* and the PNG file contains an image with no background, we can now draw the upper line of the rectangle freehand; we only have to ensure that we meet the corner points of the existing vertical and diagonal block outline. We then select one object from the *box color zone* layer, choose the purple radial gradient from the *Swatches* panel for the fill, and open the *Gradient* panel to adjust the gradient. We find the gradient displayed as a vertical guide comprising two rulers, with a white and a purple color swatch below the guide. We click the white swatch and move it to the left to adjust the gradient. Next we choose the pale tan color $C = 25\ M = 25\ Y = 40\ K = 0$ from the *Swatches* panel and use it to replace the white color in the gradient ruler of the *Gradient* panel, using *Drag & Drop*. Similarly, we replace the purple color in the *Gradient* panel with beige ($C = 35\ M = 60\ Y = 80\ K = 25$) and then using *Drag & Drop* we select the new gradient in the *Tools panel* and add it to the *Swatches* panel. In the event of problems occurring while using the illustration in a subsequent application (such as text processing, or printing it out) we can isolate and export the critical gradient elements into a bitmap file and then later replace them in the illustration, thus keeping the text and contours as vectors in the same way that we demonstrated for the transparent zones of the pie chart. We save our file under the name *etopo1_surfaceplotlight_vs7_ai_block.eps.*

As a second advanced application we will now work on the representation on the perspective of the inclined labels on the block diagram. We first incline the three labels *African Continent, Madagascar* and *Mauritius Reunion* using *Tools panel > Object > Transform > Shear*, and then type *Shear Angle* $-11°$ and *Axis Angle*

11°. Please ensure that the *Preview* checkbox is activated to permit inspection of the final result. We then shear the *Seychelles* and *Arabia* labels, choosing 59° for both *Shear Angle* and *Axis Angle*. To improve the appearance of the labels we open the *Character* panel and type 150% for the horizontal and 125% for the vertical alignment of the letters. Finally, we move the inclined labels to appropriate positions, set black strokes and text overprinting using the *Attributes* panel, and save our final graphic as *etopo1_surfaceplotlight_vs8_ai_perspective.eps*. In the event of file problems in subsequent applications due to color gradation we can replace the zone containing color gradation by exporting it as a bitmap file and again placing it in an Illustrator document, as described above for the pie chart. We then save the file as *etopo1_surfaceplotlight_vs10_ai_flattened.eps*.

8.2.7 Preparing Vector Graphics for Presentation on a Dark Background

Some users may wish to present vector graphics on a dark background for a poster or presentation, as will be demonstrated in ►Chaps. 9 and 10. To demonstrate how to modify line drawings for that purpose we open two plots, both of which have been edited in this chapter: the pie chart stored in the *icecore_piechart_vs11_ai_mask.eps* file and the block diagram in the *etopo1_surfaceplotlight_vs10_ai_flattened.eps* file. We first change all black fonts to white in both graphics and then change the black outlines in the second graphic to white. Finally, we deactivate the overprinting checkbox for white contours and fonts in the *Attributes* panel. We save versions 11 and 12 of the graphic with file names ending in ...*fordarkbackground.eps*. To prepare a third graph with dark background for inclusion in a poster or presentation we open *icecore_lineplot_vs4_ai.eps*, change all font colors to white, add a small square in the same color as the curves to link the curves with the axes, and save the file as version 8 with a file name ending in ..._*fordarkbackground.eps*.

8.3 Processing Images

In ►Sect. 7.4 we imported and processed an ASTER satellite image and saved the result as a TIFF file called *naivasha_image_vs1_matlab.tif*. We now prepare this image for inclusion in conference presentations, posters, or manuscripts. Typing whos in MATLAB indicates that the image is stored as a $4200 \times 4100 \times 3$ array consisting of three 4200×4100 arrays, one for each of the colors red, green and blue (see ►Sect. 7.4). The listing of the current variables in the workspace also provides the information that it is a *uint8* array, i.e. each array element representing one pixel contains 8-bit integers. The size of the image in the workspace is $4200 \times 4100 \times 3 \times 8$ bits $= 41,3280,000$ bits (b), or $413,280,000$ bits$/8 = 51,660,000$ bytes (B), or $51,660,000$ bytes$/1024 = 50,449$ kilobytes (kB), or $50,449$ kB$/1024 \approx 49.3$ megabytes (MB). The uncompressed baseline TIFF file created by imwrite in MATLAB is slightly larger (52.1 MB) since it includes a header providing information on the size of the image, the compression method (if any), and other properties of the image.

The satellite image processed with MATLAB is very poor in contrast, which needs to be fixed before printing. The Image Processing Toolbox in MATLAB software provides numerous methods for image enhancement, such as adaptive or non-adaptive histogram equalization using `histeq` or `adapthisteq`, respectively. Nevertheless, if the image will not be used for quantitative analysis but is to be printed out for use in fieldwork, any image processing software (such as *GIMP*, *GraphicConverter*, or *Adobe Photoshop*) can be used to quickly enhance the image.

8.3.1 Satellite Imagery: Professional Image Processing Using *Adobe Photoshop*

The *Adobe Photoshop* professional software tool included in *Adobe Creative Cloud* provides a powerful tool for editing images. Launching the program, the graphical user interface (GUI) is similar to that in the companion product *Adobe Illustrator* that we used previously, comprising the typical *Adobe Application Bar*, *Tools Panel*, and other panels. Additional information and excellent video tutorials are available from *Adobe Help* and online at

```
https://helpx.adobe.com/photoshop/topics.html
```

or

```
http://tv.adobe.com/
```

Using *Application bar > File > Open* we open the file called *naivasha_image_vs1_matlab. tif* and use *Application bar > Workspace > Photography* to choose the workspace, which shows a preselected set of panels. The *History* panel shows a list of working steps that allow us to go back and forth within the program. We adjust the image size using *Application bar > Image > Image size* to a resolution of 300 dpi and a width of 2,500 pixels.

The *Layers* panel shows our image within the layer called *background*, which is locked, as indicated by a padlock symbol. This offers the possibility of *non-destructive editing* while keeping the original image untouched. We therefore use a separate *adjustments layer* for color correction, making use of the *Curve* tool icon from the *Adjustments* panel. Clicking on the *Auto* button changes the color gradation curves for the three (RGB) color channels and creates a new layer in the *Layers* panel, with enhanced colors and contrast. We save version 5 of the file as a Photoshop file with the *.psd* extension. As files with Photoshop formats are not always compatible with later software, we also save the file as *naivasha_image_vs6_ps_rgb_2500px.jpg*, with the color profile *sRGB IEC61966-2.1* embedded. Alternative strategies for color adjustment are provided by the *auto-contrast* and *auto-color* options of the software, which can be used to enhance an image (◻Fig. 8.5) as well as for direct adjustment of the *gradation curve*.

As an advanced exercise in this subsection we will now prepare our image for offset printing, using the four CMYK ink colors. A special feature of the Adobe Creative Cloud Software is the *Color Management Workflow*, which can be controlled for all applications in the *Adobe Bridge* using *Application bar > Edit > Color Settings*. The

8

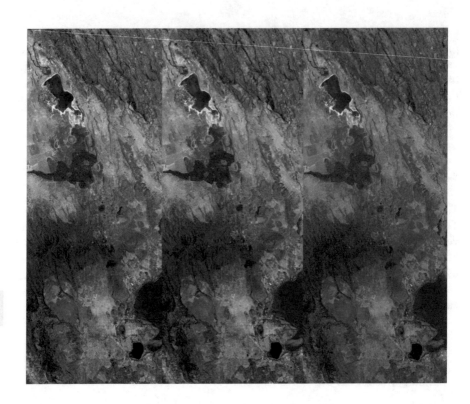

□ Fig. 8.5 Automatic color adjustments from left to right: Photoshop *auto gradation* adjustment. Photoshop *auto contrast* adjustment. Photoshop *auto color* adjustment

requirements for a perfect color management workflow include a professionally calibrated computer display, software, various technical tools, and the ability to simulate the printed outcome on the screen prior to actual printing. We continue with version 6 of the image, open the *Channels* panel, and after toggling the visibility we can see the three RGB channels used to present the image on a computer screen. As the image is intended for offset printing using the four-color process, we change the color profile into a CMYK color profile using *Application bar > Edit > Convert to profile*. In this example we embedded the *Coated FOGRA27* color profile into the file resulting in a 7.6 MB file compared to the smaller 3.3 MB RGB file of version 6 with the same image dimensions, resolution, and compression. In the appropriate dialog box we choose a suitable color profile from the *Destination Space* drop-down menu. Activating the *Preview* checkbox allows the appearance to be checked, testing the four different *Intent* options for color conversion. In the *Channels* panel we now find the four printing colors cyan, magenta, yellow, and black (also called "key"). We save the new file as *naivasha_image_vs7_ps_cmyk_2500px.jpg*, with high quality compression. More information about colors is provided in the section on *Colors and Color Management* in ►Chap. 11.

8.3.2 Georeferenced Satellite Imagery: Masking, Retouching, Adding Vectors

In ▶Sect. 7.5 we georeferenced the satellite image, i.e. the image was rotated, scaled, and deformed to fit a geographic coordinate system. We then further edited the image in the preceding subsection of this chapter by auto-adjusting the colors and contrast and then converting the color model from RGB to CMYK. The resulting raster image, which has a black background, shows a raster coordinate system and labels. Our task in this subsection will be to enhance the colors and contrast, to remove the background, and to add a vector coordinate system, labels, scales, and vector objects such as tectonic faultlines and the outlines of waterbodies. The original georeferenced version in *naivasha_georef_vs1_matlab.png*, generated in MATLAB, is 4,500 pixels wide resulting in a width of 19.05 cm at 600 dpi resolution. To reduce the file size we choose *Application bar > Image Size*, where we can set the *Resolution* to 300 dpi and reduce the *Width* to 2,500 pixels. Activating the *Chain* symbol ensures that we keep the original proportions of the image. We now save the 2.2 MB file as *naivasha_georef_vs3_ps_rgb_2500px.tif*.

To select and remove the black background of the image we now choose the *Lasso Marquee* tool and click on the four corner points of the colored area in order to make a selection around the edge of the satellite image. We can press the *Alt* key to reduce the size of the selected area, or the *Shift* key to increase it. To select the black zone we then invert our selection using *Application bar > Select > Inverse*. Using the *Backspace* key opens the *Content* dialog window and we choose to replace black with white. To control the selection we can always switch view between the *Standard* mode and the *Mask* mode by toggling the lowermost tool in the toolbox. In *Standard* mode the borders of the selection are marked by an animated border resembling crawling ants. In the *Mask* mode the selection is indicated by a red semi-transparent zone overlying the image. In case we want to reload the selection later, we can save the mask as an *alpha channel* by using *Application bar > Select > Save Selection*. In order to retain the alpha channel we save the image as a Photoshop file named version 3, because the JPEG file format does not allow us to save separate layers. We can also save the image as a TIFF file named *naivasha_georef_vs4_ps_rgb_white.tif*. When adjusting the image colors (with either a black or a white background) using *Application bar > Image > Adjustments > Curves > Auto*, the results are much less attractive in terms of color intensity, brightness, and contrast than in the non-georeferenced version. This may be a result of the process taking into account the black or white background.

For a georeferenced image combined with a coordinate system, we open *naivasha_georef_vs2_matlab.tif* and try to again adjust the colors using *Application bar > Image > Adjustments > Curves > Auto*. For some reason this has almost no effect on the colors. We could of course just copy and paste the version 3 image generated above. Let us assume, however, that we wish to use the non-georeferenced image stored in the file called *naivasha_image_vs6_ps_rgb_2500px.tif* as it has nice colors, and that we want to georeference this image using *Photoshop*. If the rulers are not visible in the document window we can make them active using *View > Rulers*. By clicking and dragging the mouse down from the horizontal *Ruler* we create two horizontal *Guides* and, in a similar manner, two vertical *Guides* from the vertical ruler. With the aid of the *Guides* we mark the four corner points of the satellite image. We now copy and paste the image from *naivasha_image_vs6_ps_rgb_2500px.tif* into a new layer at the top of the layer list and can then observe the image while we transform it into a trapezoid using *Application bar > Transform > Distort*. Finally, we edit the black background behind the image by drawing a white rectangle between the guides in a layer between the two images, and we then save the file in the Photoshop file format as version 5, or alternatively as a JEPG file called *naivasha_georef_vs6_ps_rgb_coord.tif*.

To demonstrate how to add vector objects in Photoshop we now add text and some contours using the *Pen* or the *Line Drawing Tool* and the *Horizontal Type Tool* from the Tools panel. The tools look very similar to the Illustrator tools and we can again organize all different types of elements into separate layers. To demonstrate the limitations in compatibility between different modules of Adobe Creative Cloud, we open and then copy some random line-drawing elements. For example, we can select and copy the *fluvial system* layer included in the file named *srtm_faults_sat_merged_vs1_ai.eps* from Illustrator onto the computer clipboard, paste it into our Photoshop file as a *Smart Object*, and adjust its dimensions. Unfortunately, before we can continue with our work the program tells us that all objects have now been converted into a pixel format, as indicated by the message *Rendering Placed Document*.

For text and line drawings we prefer to preserve editable vector text, line drawings, and elements rather than having them rasterized by *Photoshop*. To retain vectors we therefore follow a different strategy and go back to using *Illustrator*. We reopen the vector file *srtm_faults_sat_merged_vs1_ai.eps* using *Application bar > File > Place…*, import the raster image file called *naivasha_georef_vs6_ps_rgb_coord.tif* into the *sat coordinates* layer, and adjust its size and position there to create congruency between both coordinate systems. We can type labels for *Lake Naivasha, Mau Escarpment* and *Aberdare Ranges* in the uppermost layer, as we did with Photoshop. When saving the illustration we observe the file size and discover that the *.eps* vector file that was originally 1.2 MB becomes a very large file when combined with the original georeferenced raster image. A *.ai* file with the option to embed a PDF deactivated will have a smaller file size than the *.eps* file, but for reasons of compatibility we save it as a *.pdf* file. To remove the white background of the image and reduce the file size, we replace the image with a version that has no background called *naivasha_georef_vs9_ps_cmyk_670px.pdf*, generated in Photoshop, and finally save our illustration as *srtm_faults_sat_merged_vs3_ai.pdf*.

8.4 Editing Text

This section is on preparing text for integration into manuscripts, posters and, less commonly, into presentations. Word processing is one of the oldest applications of computers and the emergence of sophisticated software tools that were originally independent of each other, each of which subsequently claimed to be market leaders, has led to a wide variety of text formats and consequent incompatibilities between operating systems and software tools. The *Guidelines for Authors* of most journals therefore recommend that the layout of a manuscript be kept simple, avoiding the software's options to justify text or hyphenate words, and that the manuscript instead be submitted as single-column left-justified text.

The text format with the highest possible compatibility is the *American Standard Code for Information Interchange* (ASCII) format, which was first published in 1963 by the American Standards Association (ASA). As a 7-bit code, ASCII consists of $2^7 = 128$ characters (codes 0 to 127). Whereas ASCII-1963 lacked lower-case letters, these were subsequently included in the ASCII-1967 update, as well as various control characters such as *escape* and *line feed*, and various symbols such as brackets and mathematical operators. Since then a number of variants have appeared to facilitate the exchange of text written in non-English languages, such as the expanded ASCII, which contains 255 codes (e.g., the Latin-1 encoding).

One of the classic examples of cross-operating system or software incompatibilities is the *carriage return/line feed* problem. *Carriage return* (CR) in computing refers back to the return of the carriage on a typewriter and is an ASCII control character to move the cursor back to the first column, or position, on the same line. *Line feed* (LF) moves the cursor to the next line on a typewriter, or starts a new line on a computer. Both are used together to mark the end of a paragraph and the start of a new one in most word processors. Unfortunately, some software tools use only the carriage return character to mark the end of a paragraph and some only the return character, while other software tools use carriage return together with line feed. The problem is that text generated on UNIX-based operating systems (such as Linux, SUN Solaris, or macOS) using LF alone as end marks for paragraphs appear as a single long line on Windows computers. On the other hand, Windows-sourced files with both CR and LF appear with a second paragraph break, and hence with an extra line between paragraphs. Although most word processors can now deal with this problem, users still come across the CR/LF problem when working with low-level word processors such as those used for coding software.

The more advanced and feature-rich the word processor is, the more incompatibilities appear when exchanging text between different computer platforms or software. In 1987 the Microsoft Corporation introduced the cross-platform 8-bit *Rich Text Format* (RTF), which includes *escape sequences* allowing the text to be formatted using a *WYSIWYG* (*what you see is what you get*) type text processor (►http://microsoft.com). Binary DOC and DOCX files generated by *Microsoft Word*, ODT files generated by *OpenOffice Text*, and *Pages* files generated by *Apple Pages*, all contain a great deal more formatting information than RTF files, but at the expense of being less compatible. *Apple Pages*, however, contains converters to read DOC and DOCX files and convert them into *Pages* files, but the latest 2016 release of Microsoft Word can not read *Pages* files.

The use of a word processing program is comparatively simple. For most representatives of the older generation of scientists who grew up with typewriters, long before the era of the World Wide Web, digital photography and social media, it was usually the first application that they came across. It is interesting to note that, in our experience, basic functions of modern word processing are often ignored, in particular the use of *paragraph styles* and *character styles*. This is mainly due to the fact that a text can be easily formatted without these functions. For example, you can mark parts of the text and change font types and sizes, define the type alignment, and make individual words bold without having previously defined styles. However, the use of paragraph and character styles is very helpful when you want to change styles quickly in longer texts. For example, if you want to quickly change the font size of all headings from 12 to 14 points or put all fossil names into italics, then it is a big advantage if the paragraphs and individual characters have been previously tagged. Typesetting tags are used to define paragraph and character styles, each of which is associated with specific text-formatting attributes such as font type and size, text alignment, color, and many more. In addition to these obvious advantages of using paragraph and character styles, the preparation of tagged text documents is a prerequisite for the submission of manuscripts to most publishers. It makes it easier for you to work and thus reduces your costs. Typesetting is often an automated process, i.e. it is carried out by machines. Automated, unattended typesetting is criticized by professional typesetters because of the high error rates involved in these procedures, while the publishers appreciate the rapid turnaround times and increased cost efficiency.

When formatting a text in a text editor such as *Word*, *Pages* or *OpenOffice*, the application generates a code showing a typical syntax, similar to that used in other programming languages. Well-known examples of coding, tagging coding, and tagging texts are included in the HTML or XHTML mark-up languages that are used to create webpages. When writing a piece of text, all text elements are tagged using a structure such as *header1*, *header2*, *header3*, *normal*, *indent*, *table*, or *list*, which enables the use of *styles* instead of directly formatting each individual character and paragraph of a manuscript. The characteristics of a *paragraph style*, such as its *font*, *font weight*, *font size*, *color*, etc., are defined in a central document called a *style sheet* and can be administrated therein. The formatting of individual characters, individual letters, or parts of text within a paragraph, is defined in the *character style*, which is then available throughout the entire document. For websites this document is called a *css* or *cascading style sheet*, which is used to improve the appearance of the site. Storing all style information in a central place allows different styles to be applied and, if necessary, changed quickly, consistently, and safely throughout the entire document. The easiest way to get started with the use of styles is to adopt a ready-to-use template, such as may be provided by the program itself or downloaded from either the authors' guidelines of the journal in which the document is to be published or the graphics design department of the institution supporting the publication.

Starting *Microsoft Word 2016* on a computer running *macOS* opens a document browser with ready-to-use templates for many purposes. We choose *File > New Document*, which opens a window with a new blanc document, together with a toolbar with several tabs such as *Home*, *Insert*, *Design*, and *Layout*. Choosing *Word > Preferences* from the menu opens a window in which we can personalize the program, for example by changing the *User Information* and *AutoCorrect* settings. We type some placeholder text (for example: *Heading 1, Heading 2, Heading 3, Text Hello World*) and format it using the paragraph styles with the same names from the *Styles Panel* on the *Home* tab. If we need some more text, we search the web for any *blindtext* generator and then paste some 200 words as running text below each heading. Alternatively, we can open the ASCII file *newpaper_vs1.txt* available online through Springer Extras, which contains a header, the list of the authors and their affiliations, headers such aa *Abstract*, *Introduction* and *Regional Setting*, and a short piece of blindtext starting with *Loren ipsum*:

```
This is the title of our manuscript
John Q. Scientist, Jane W. Researcher
University of Nowhere, Department of Science
```

Abstract
Lorem ipsum dolor sit amet, consectetur adipisici elit, sed eiusmod tempor
incidunt ut labore et dolore magna aliqua. Ut enim ad minim veniam, quis
nostrud exercitation ullamco laboris nisi ut aliquid ex ea commodi
consequat. Quis aute iure reprehenderit in voluptate velit esse cillum
dolore eu fugiat nulla pariatur. Excepteur sint obcaecat cupiditat
non proident, sunt in culpa qui officia deserunt mollit anim id est laborum.

We click within the individual paragraphs and apply a style to these paragraphs, e.g. by selecting the style *Title* from the *Styles Panel* after positioning the cursor within the first paragraph. Hovering over a style in the *Styles Pane* shows the format details, i.e. the *Font, Font Color, Space, Numbering* and *Theme*. On the *Design* tab we can choose different themes, which are built-in sets of fonts and colors. The *Office* theme is suitable for a scientific report, using the non-serif font *Calibri* for the main text and the headings. To make a single word italic within the first paragraph, in our example the title, we first mark the word (e.g. *manuscript*). In Word, paragraph and character styles appear in the same list in the *Styles Panel*. There, they are distinguished by a blue symbol to the right of the name of the style: a paragraph mark (¶) for paragraph styles and a small symbol (a) for character styles. After marking the word, or even individual letters, we can apply a predefined character style by clicking the name in the *Styles Panel*. Alternatively, we can change the style of individual characters (e.g. to bold) and define a new style by clicking on *New Style* in the Styles Panel. Changing the style is achieved by selecting an individual paragraph or character, changing the style using the formatting tools, pointing on the blue symbol to the right of the style name until a small black triangle appears, and choosing *Update to Match Selection*. This process changes the style throughout the text.

The *Home* tab includes various formatting options for the text. To increase efficiency and to maintain an overview of invisible elements in a document, we choose *Show all nonprinting characters* on the *Home* tab, which shows *page breaks, paragraph marks* or *pilcrows* (¶), and *soft returns* or *manual line breaks* (↵). Each end of a paragraph requires a paragraph mark, which also embodies the paragraph style. Finally, having formatted the table of contents we save the file as a *Word Document* and also, for future use, as a *Word Template*. When opening a text from another computer the formatting may look a little strange if some of the fonts are not available.

The workflow for editing the same document in *Apple Pages* 2016 is very similar to that in Word. Opening *Pages* for the first time shows a welcome window and a brief getting started video tutorial at

http://apple.com/iwork/

followed by the *Template Chooser* window with templates for word processing and page layout. We explore the *Application* bar and then open the *MS Word* file *report_word_nonserif.docx*. To customize our workspace we choose *Application bar > View > ...* where we can make visible *Invisibles, Comments, Rulers*, and also the main tool for formatting, the *Inspector* window. In the *Format* options on the *Tools* panel we find a list of the same styles as were defined in Microsoft Word, as well as a preset styles offered by *Pages*. To demonstrate how changes can be made very quickly, intuitively and consistently, we place the cursor within *1. Heading 1* in our text, open the *Colors* window, and change the color to black for some letters of *Heading 1*. After clicking on the small gray triangle in the *Paragraph Styles*, we first choose the *Heading 1* style and then *Redefine from Selection*.

As an example of adding a figure, we simply click on *Application bar > Insert > Choose* and import *icecore_piechart_vs11_ai_mask.eps*, the pie chart edited in ▸Sect. 8.2. Using *Application bar > Format > Image > Mask with Shape > Rectangle* we can crop the white space around the image and adjust its size. Finally, we save it as *report_pages_nonserif.pages*.

8.5   Editing Tables

In this section we deal with preparing tables for integration into manuscripts and, less commonly, into posters. Tables should not generally be included in presentations. As with word processing, tables were an early application of computers since the first computers were designed to process data rather than text, and data are usually stored in tables with rows and columns.

In ▸Chaps. 4–7 we used MATLAB to import, process, and visualize data organized as arrays of numerical values, as well as text and other types of data. We have not used Microsoft Excel spreadsheets even though MATLAB can read its files using *xlsread*. Instead, we mostly used *load* to import data from ASCII files and in some rare cases we also used other tools to read specific binary formats such as images, or digital elevation models. The reason for using ASCII instead of Microsoft Excel files is to avoid the incompatibilities that we often come across when working with binaries. Again, the *Guidelines for Authors* for most journals recommend that the layout of tables attached to a manuscript be kept simple. Both tables and figures are generally submitted as appendices to a manuscript, rather than integrated into the text.

Tables are typically two-dimensional rectangular arrays of numbers or text, organized in rows and columns. In contrast to data arrays to be imported into MATLAB, tables to be included in printed manuscripts are restricted in their dimensions by the page format. The widths of the tables in this book, for example, are limited to the 11.69 cm width of the text column. As an example, we use the text file called *geochem_data.txt* that was saved as an ASCII text file in ▸Chap. 4.

```
SampleID    Percent C       Percent S
101           0.3657          0.0636
102           0.2208          0.1135
103           0.5353          0.5191
104           0.5009          0.5216
105           0.5415         -999
106           0.501          -999
```

This file is in a very common format for tables downloaded from internet resources. The first row contains the names of the variables and the columns provide the data for each sample, i.e. the sample identifications and the percentage of carbon and sulfur in each sample. The absurd value −999 indicates missing data in the data set. The columns of tables are separated by *delimiter characters*, commonly single spaces, multiple spaces, or tabs. In some examples commas, colons or vertical bars are also used as delimiters. The lines are separated by *newlines*, or *line breaks*, or *end-of-line* (EOL) markers, which in ASCII character sets are commonly CR, LF or CR/LF (see ▶Sect. 8.4).

Taking a closer look at our example file you may notice that the delimiters of the values are multiple spaces, which is the worst-case scenario for desktop publishing. We can *unhide* the control character used by our word processing software in order to reveal the spaces:

```
SampleID···Percent·C········Percent·S
101···········0.3657···········0.0636
102···········0.2208···········0.1135
103···········0.5353···········0.5191
104···········0.5009···········0.5216
105···········0.5415··········-999
106···········0.501···········-999
```

Formatting such a table is a nightmare and we therefore need to replace the spaces with tabs. Tabs or tabulator stops on a typewriter halt the carriage movement by mechanical gears and allow text to be aligned vertically. In word processing, the same idea is used to align text with the left and/or right margins, to centralize it, or to arrange numerical values in such a way that the decimal points in figures on separate lines are positioned directly beneath each other.

There are a number of different ways to replace the multiple spaces with single tabs. One way is to first replace double spaces with single spaces using the *search-and-replace* feature of the word processor. We use this feature repeatedly until no more double spaces are found.

```
SampleID·Percent·C·Percent·S
101·0.3657·0.0636
102·0.2208·0.1135
103·0.5353·0.5191
104·0.5009·0.5216
105·0.5415·-999
106·0.501·-999
```

We can then replace single spaces with tabs, again using *search-and-replace*. The control character for a tab is a small arrow. Some of the tabs mistakenly inserted in the header have to be changed back to spaces.

```
SampleID·Percent·C·Percent·S
101·0.3657·0.0636
102·0.2208·0.1135
103·0.5353·0.5191
104·0.5009·0.5216
105·0.5415·-999
106·0.501·-999
```

We can then change the location of the tab stops to alter the layout of the table in the word processor, e.g., by introducing two left tab stops, at 3 and 6 cm.

```
SampleID→Percent·C→Percent·S
101→0.3657→0.0636
102→0.2208→0.1135
103→0.5353→0.5191
104→0.5009→0.5216
105→0.5415→-999
106→0.501→-999
```

We can now convert the tab-delimited text into a formatted table, either using the word processor's *table* feature or by importing the text into spreadsheet software such as Microsoft Excel, OpenOffice Spreadsheet or Apple Numbers. Alternatively, we can import the ASCII text file delimited by multiple spaces into one of the software tools that allow multiple spaces to be replaced directly by single tab delimiters. The binary *xls* or *.xlsx* files generated by Microsoft Excel, *.ods* files generated by OpenOffice Spreadsheet, or *.numbers* files generated by Apple Numbers, contain a great deal more formatting information than simple text files, but at the expenses of being less compatible.

The simplest way to import the contents of a tab-delimited table into a text editor or professional *desktop publishing* (DTP) program is to create a new table in the document and then to select the text and tabs and just *copy and paste* them as a whole into the selected cells. Available formatting options include the possibility of hiding or using visible *cell borders* and choosing the type of fill for the cells in the table (with a choice between different colors, alternating colors, gray or *no fill*) for the table cells. Furthermore we can change the appearance of *headers* or *footers* and adjust the *inset margins* and *text alignment*.

Recommended Reading

Alley RB (2000) The Younger Dryas cold interval as viewed from central Greenland. Quat Sci Rev 19(1–5):213–226

Trauth MH (2010) MATLAB® Recipes for Earth Sciences–3rd Edition. Springer, Berlin

Creating Conference Presentations

Electronic Supplementary Material The online version of this chapter
(https://doi.org/10.1007/978-3-662-56203-1_9) contains supplementary material,
which is available to authorized users.

9.1 **Introduction**

The results of a project are typically presented in three formats: posters, talks, and papers. A *poster* is collection of graphics, photos, and text printed on a large sheet of paper that is presented on a poster-board in a large hall in a conference building. Before the actual presentation of a poster, the scientist submits an *abstract* summarizing the key findings of the research project. Abstracts are typically limited to between 100 and 200 words although exceptions are sometimes made, e.g., for extended abstracts which, in some cases, may reach the length of a journal article. Abstracts are also required to be submitted for a conference *talk*. A talk is an oral presentation, typically 15–20 min long and supported by a series of pages or slides projected by a video projector. The slides generally contain graphics and photos, but only a relatively small amount of text. Talks are organized into theme sessions that are chaired by a session convener. The convener introduces the presenter of the talk, takes care of the time management of the session, and moderates the discussion with questions and answers after the talk. This chapter deals with the planning of a talk and the design of presentation slides, as well as offering a few suggestions for practicing and presenting a talk.

9.2 **Planning an Oral Presentation**

Numerous books, articles and videos already exist on planning an oral presentation, designing slides, and delivering a talk. Excellent books on oral presentations have been published by Nancy Duarte at *Duarte Design Inc.* (► http://duarte.com) (Duarte 2009, 2010) and Garr Reynolds (► http://garrreynolds.com) (Reynolds 2010, 2011, 2013). Garr Reynolds also has a blog called *Presentation Zen* that provides comments on numerous examples of great presentations given by other people (► http://presentationzen.com). An important reference for data visualization in science is the work of Edward Tufte (► http://www.edwardtufte.com). One of the best video examples of presentations is of Steve Jobs, the late CEO of Apple Inc., presenting the first *iPhone* in the Macworld San Francisco 2007 Conference's Keynote Address, available from

```
http://itunes.apple.com/us/podcast/apple-keynotes/id275834665
```

and from several other websites, including *YouTube* (►http://youtube.com). In this video you can see the presentation of the iPhone, which starts at 26:20 min into the video and continues for most of the remaining keynote address. The first four minutes or so are actually structured like a scientific presentation. The first sentence, "Every once in a while a revolutionary product comes along that changes everything", clearly builds up a high level of expectation, in particular as there were many rumors about the new mobile phone before its actual launch. In contrast to a scientific presentation, however, Jobs then goes through some of Apple's major achievements and products, rather than discussing the problems with current mobile phones, before then revealing to the audience which new product he is going to talk about over the next hour or so.

At 29:18 min he presents the new product by simply displaying its name on a slide (◻Fig. 9.1). In a science talk or in a paper, this would correspond to the *here we present* statement. The statement that "Today, Apple is going to reinvent the phone" clearly

🔲 Fig. 9.1 Steve Jobs introducing the new iPhone by simply displaying its name on a slide. The sentence, "Apple reinvents the phone", which is written on the slide, clearly indicates the importance of the new product being introduced (Screenshot from the keynote address at the Macworld San Francisco Conference on 1st January 2007, by Apple Inc.)

indicates the importance of the new product being introduced. Then at 29:48 min Jobs presents the current *state-of-the-art* of mobile phones which, in a scientific presentation, would have been outlined before presenting the new product (🔲Fig. 9.2). He does this by explaining a very simple *xy* plot with an *easy-to-use* axis and a *smart* axis. After explaining the axes of the empty graph, he places colored circles on the graph, one by one, representing some of the smartphones offered by competitors, before eventually placing the iPhone in the best possible position on the plot of *smart* versus *easy-to-use*. At 30:58 min, he repeats the "Today, Apple is going to reinvent the phone" statement and concludes his introduction of the keynote address before then continuing with an explanation of the technical details and features of the new product for the next hour.

Steve Jobs' presentations have been widely discussed on the World Wide Web by many people, including Garr Reynolds. One of the most striking features of his presentations was the simplicity of the slides, with only a single statement or message per slide. Most slides comprised a single photo of an item such as a mobile phone, one word such as iPhone, or a very simple graphic. If you go through the entire keynote address, you will find that most of the presentation is made up of still or animated photos of the phone. If text is used then it is reduced to an absolute minimum, such as a short list of the features of a new product, in which all features are described by one or two words only, such as at 1:32:06 h when the typical features of smartphones are listed. Instead of also listing the features of the iPhone as text, he quickly goes through a series of photos of the phone, showing the features as he talks about them. Then at 1:32:40 h he presents the two available models by showing photos of the iPhones in the middle, with text indicating the size of the flash memory to the left and the price to the right. Tables are rarely used but are necessary when comparing the features of different products. The tables then have very few rows and columns, perhaps comparing only

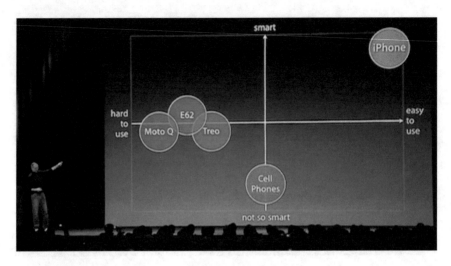

▪ Fig. 9.2 Steve Jobs presents the state-of-the-art for mobile phones, which would be outlined before presenting the new product if it were a scientific presentation. He does this by explaining a very simple *xy* plot with an *easy-to-use* axis and a *smart* axis. After explaining the axes of the empty graph, he places colored circles on the graph, one by one, representing some of the smartphones offered by competitors, before eventually placing the iPhone in the best possible position on the plot of *smart* versus *easy-to-use*. (Screenshot from the keynote address at the Macworld San Francisco Conference on 1st January 2007, by Apple Inc.)

three to five features in just two products. A good example of how to put together a graphic in a presentation can be found at the end of the presentation (1:41:15 h). Jobs again uses animation to slowly put together the various elements of bar plot, rather than straight away displaying the completed plot.

A second point to note in this presentation is the way that Jobs communicated with the audience. In contrast to other presenters, Jobs never showed too much excitement or agitation in his presentations, but he was clearly enthusiastic about the latest Apple products. He also made it very clear several times during his presentations that he worked with a great team. He maintained eye contact with the audience but also admired the products that he and his team had created over recent months and years, when they appeared on the slides. Jokes are well measured during his presentations, for instance when presenting a photo showing an old iPod with a rotary dial as the new iPhone at minute 29:30 and waiting for the audience to react.

The third thing to note in this presentation is the way things are said. It is well known that Steve Jobs practiced his presentations many times until it looked effortless. Large numbers of backstage staff were involved, as well as a lot of technology, to ensure that the keynote address ran smoothly and without any major incidents. There are some rare examples of problems during Jobs' presentations and it is interesting to observe how he handled the problems. During this presentation, he speaks slowly, makes pauses, varies the speed at which he speaks, raises and lowers his voice, and repeats things to emphasize their importance. After 1:45:20 h of the presentation he wraps up by again displaying some of the most successful Apple products followed by the new iPhone, and then repeating a quote from Wayne Gretzky, showing a historical photo of the Apple Computers company (now Apple Inc.) in its early days, thanking the audience, and concluding the presentation.

Garr Reynolds, who worked for Apple in the past as a Manager of their Worldwide User Group Relations, provides a second example of a good presentation. As a presenter he shows a lot more enthusiasm, makes many jokes, and communicates directly with his audience from the very first minute. In a lecture Reynolds gave at the 2009 Citrix Synergy conference in Las Vegas, he starts with "Hello Las Vegas", followed by a charming laugh and "How are you doing? Are you alright?". He then continues by politely asking the audience whether they got enough sleep last night, and asks them to respond by raising their hands or their wine glasses. The first twelve seconds of the presentation aims to establish contact with the audience, to grab their attention, and to get people involved in the show, making very sure that they know that the presenter will continue to demand responses throughout the rest of the presentation. There is no escape from this highly interactive lecture!

It is strongly recommended that you watch some of Reynolds' lectures to see how the slides are designed, and how Reynolds stays in touch with his audience. Garr Reynolds believes in simplicity in slides, in the overall message, and in the way things are presented. He aims for high signal-to-noise ratios, such as in the simple xy plot that Steve Jobs' used to describe smartphones in terms of how *easy-to-use* and *smart* they are, using it to explain that while software technology is growing more and more sophisticated it is also expected to remain simple to use. While the graph remains on display Reynolds explains the plot several times, each time using different wording in order to make it quite clear to the audience.

The key message of simplicity in Reynolds' presentation appears after about 20 min during the 2009 Citrix Synergy talk when a slide is shown with a photo of a delicately balanced stack of rounded pebbles on the right and a quote on the left from the late artist, designer, and architect Koichi Kawana: "Simplicity means the achievement of maximum effect with minimum means". The slide itself is very simple, despite including text. The text is, however, given additional structure a few seconds after the slide is displayed by simply adding color to the second part of the sentence, "… maximum effect with minimum means". A very effective way to add structure to text in presentations, as well as on posters and in brochures, is to highlight keywords by making them bold, underlining them, or adding color. The presentation then provides numerous other examples of simplicity in slides, contrasted with examples of bad (complex) slides and, after minute 39:20, the example of Al Gore becoming a better presenter following training by Duarte Design Inc.

Nancy Duarte is the third example of a good presenter. She provides numerous video lectures and examples of presentation slides on her webpage, which also includes a blog. Her presentations are less sober than those given by Reynolds, more like the sort of presentations that you might expect from an artist or, if you like, more feminine. Good examples of video lectures can be found on her webpage under ►http://duarte.com/.

There is no problem at all with also using this more light hearted design of slides in a scientific presentation. I remember very well a presentation at the 1992 Annual Meeting of the Geological Society of America (GSA, ►http://geosociety.org), when Kenneth M. Schopf used the example of Mikado pick-up sticks to explain ecological locking and the stability of fossil morphologies (Schopf et al. 1992). During his presentation photos were shown in which sticks were picked up, provoking movement and reorganization of the remaining sticks, representing the changing interactions between organisms of the ecosystem. Whereas the scientific topic was very theoretical and difficult to understand, the presenter kept the audience attentive by showing great photos of Mikado sticks from time to time. Similarly, Axel Meyer, biologist at the University of Constance in Southern Germany, presenting at a symposium on the East African Lakes in Jinja, Uganda in 1993, broke up his talk on using the molecular-clock approach to dating the dispersal of East African cichlids by showing colorful fish on every third or fourth slide. More artistic slides using the hand-drawn effect of graphics or presentation software may have the same effect, making the presentation look more alive than a more sober presentation design. During her presentations Duarte also communicates with her audience, asks them questions, and maintains eye contact with individuals in the audience.

The way oral presentations are given very much depends on the discipline, the topic presented, and the audience attending the presentation, as well as reflecting the personal preference of the presenter (as in some of the examples provided). However, some fundamental rules that are independent of the discipline, topic, and audience apply to oral presentations and need to be kept in mind when planning such a presentation. The planning of a presentation includes the following three steps:

1. Deciding what message you want to deliver, and to whom. – Decide on the main points of your presentation, avoid unnecessary detail, and keep to a minimum the quantity of information to be delivered. The way the presentation is designed will depend on the individuals making up the anticipated audience, their levels of relevant knowledge, their interests, and their expectations.

2. Designing the concept and didactics of the presentation. – A good presentation has a clear and simple structure that will need to be defined before deciding on the contents of the slides. Paper cards can serve as analog slides that can be rearranged on a table when going through the concept of the presentation, prior to going digital. A template should then be created and a unified layout selected for all slides, including colors and fonts, before actually creating the individual slides.

3. Practicing and delivering the presentation. – Most people write down the text for a talk in order to better plan the presentation and to estimate its duration. If you do so, remember that memorized written text never works in conferences. A lot of practice is necessary instead, in order to become detached from the written text and to ensure a truly live presentation.

The following sections describe the second and third of these three steps in greater detail.

9.3 Designing the Concept

As an example, let us design a two-minute presentation. In a course held at the University of Potsdam on collecting, processing, and presenting geoscientific information, students organized one-hour sessions with 10–15 two-minute presentations, chaired by the project leaders and their deputies. A large lecture hall provided a great conference-type atmosphere with its audio-video system, artificial-light dimming and window shading systems, and large projection screens. Two minutes may seem far too short a time in which to present anything, but experience from the course clearly shows that this is an ideal length of time for a talk by a beginner.

Assuming that we have simple slides with no text, three to five slides can be shown during such an oral presentation, depending on the complexity of any graphics included in the talk. Much has already been written on the best concept for a series of slides.

The classic idea for presentations, which has been stated numerous times elsewhere, is that you should "Tell them what you're going to tell them, tell them, and then tell them what you told them". Of course this is a good concept in general, but many talks have been rendered quite boring because presenters have adhered too strictly to this adage, and the talk that stands out in your memory is often the one in which the presenter dared to do things differently. We will come back to this point later when we see that a good presenter may indeed follow this golden rule, but in a way that is not always obvious to the audience. Boring examples of presentations, however, typically have the following structure:

1. The first slide shows a long title, many authors, the affiliations of the researchers, and the logos of any sponsors. While showing this slide, the presenter typically starts by thanking the co-authors and sponsors, and then explains the basic idea of the project and how it all started.

2. The next slide presents the outline of the talk, usually as a bullet point list, and although everybody knows what to expect the presenter tells the audience that he or she is going to present some background information on the project, the methods, the results, and an interpretation or discussion of the results, followed by a summary and an outlook for future research.

3. In the presentation that follows the presenter sticks to the outline and presents the results and interpretation, slide by slide. These include many text slides, bullet point lists, large tables, and complex graphs with small fonts and thin lines.

4. The most boring part comes at the end, when slides are shown that summarize the talk. In some rare cases the presenter, realizing that it would be very boring to read out all of the text, has been known to simply invite the audience to read it for themselves. In most cases, however, the presenter really does read it all out, as if the presentation would not be complete without reading the summary.

You may, however, have noticed that none of the presenters discussed in the previous section followed this type of structure. Steve Jobs says things like "Every once in a while a revolutionary product comes along that changes everything" and then "Today, Apple is going to reinvent the phone". Later, he says that he will introduce an iPod, a phone and an internet device rolled together into a single unit. This indeed follows the rule to tell the audience what he is going to talk about. The difference, however, is that he does not explain the concept of the talk in the way that many people would do. Instead, he tries to attract attention and to increase anticipation by promising something excit-ing (a new device), before then going on to outline the new product, rather than just presenting an outline of the talk. Garr Reynolds does the same thing, spending a lot of time introducing the topic of the talk rather than introducing the talk itself, after hav-ing first established contact with the audience through his opening remarks.

In essence therefore, by all means tell your audience what you are going to talk about, but not how you are going to do it. Your colleagues certainly know already that you will show them your sample locations on a map, the equipment that you used to analyze the samples, graphical presentations of the results, and finally, your interpreta-tion of the results.

To start a two-minute presentation on Neil Robert's surprising results of a dry Younger Dryas in East Africa (from ▶Chap. 2) you could, for example, show as the first slide a beautiful photo of Lake Magadi and the text:

```
The Younger Dryas in Africa
Neil Roberts
University of Leicester

together with
Maurice  Taieb,  Philip  Barker,  Brahim  Damnati,  Michel  Icole  and  David
Williamson
```

using the largest font for the title and a small font for the co-authors. If you prefer, you could also add a subtitle, such as

```
The Younger Dryas in Africa
Wetter, drier, or no change at all?
```

We criticized the title of the paper for not including any information on the way that the East African climate changed during the Younger Dryas. The title of the talk, however, could deliberately exclude this information in order to increase interest and to make the audience curious about the outcome of the investigations in the Magadi basin. Audience curiosity can be stimulated by conference presentation titles such as "Life on Mars", with people attending because they want to know if the title is a question or an answer, and whether the researchers have indeed found evidence for life on the red planet.

While the slide is displayed you could explain to an audience that is not familiar with the current discussion on the influence that the Younger Dryas had on tropical climates that contradictory results had been obtained from previous investigations completed in various locations at lower latitudes. Nobody, however, had ever looked at paleoclimate records from East Africa, which was why Roberts went to Lake Magadi and collected a sediment core, so that he could look at the Younger Dryas interval in greater detail. The next slides would then follow a logical sequence, starting with a map showing the location of Lake Magadi and the sediment core that Roberts col-lected, then a photo of the core cut in half, and then the climate proxy record and age dates showing that the lake's water level was low during the Younger Dryas, suggesting a drier climate. If you wish to convince the audience of your story it is best to avoid any unresolved questions. On the other hand, if you expect to receive advice from the audience, consider your questions carefully. To wrap up you could say that, at least in

East Africa, the Younger Dryas indeed seems to have caused an increase in aridity. After two minutes, you thank the audience and the chairperson asks for questions.

9.4 Creating a Template

Having prepared an analog draft of the presentation, as suggested by Garr Reynolds, we now launch the presentation software (such as *OpenOffice Impress*, *Microsoft PowerPoint*, or *Apple Keynote*). We first create a template for the slides in our presentation that defines their size, background, color schemes, animations, font types, and font sizes, as well as other layout elements such as logos, slide numbers, and so forth (◻Fig. 9.3). All presentation software tools come with a gallery of design templates. Most of these, however, have been seen many times at conferences and should therefore not be used, except as an inspiration for our own template.

We then decide on the size of the slides, remembering that almost all presentations are in landscape orientation. Conference presentations are typically displayed on a screen using a digital projector. Projectors vary in both their resolution and their brightness. Common projector *resolutions* are Super Video Graphics Array (SVGA, 800 × 600 pixels), Extended Graphics Array (XGA, 1,024 × 768 pixels), and the two High Definition (HD) formats 720p (1,280 × 720 pixels) and 1080p (1,920 × 1,080 pixels). If you are not sure what the resolution of the projector to be used at the conference will be, and especially if you expect it to be an older device, choose an XGA resolution to be on the safe side. A few of the lower-cost or older devices that are used at workshops can only handle SVGA resolutions, but modern

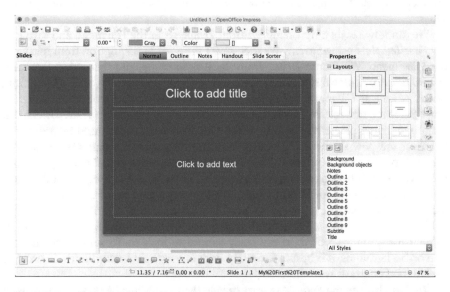

◻ **Fig. 9.3** Screenshot of the graphical user interface of *OpenOffice Impress*, including the *Slides* panel (on the left), the presentation slide itself (in the center), and the *Tasks* panel (on the right). This slide displays the template used for our presentation, with a blue background and two text boxes

high-definition projectors are normally used at larger conferences rather than SVGA or XGA projectors.

Most modern projectors have a *brightness* of 1,500–2,500 ANSI lumen, which is suitable for small workshops or lectures, while more expensive projectors with a brightness of more than 4,000 ANSI lumen are used in large conference or lecture halls. The brightness of the projectors is an important aspect to consider when creating a template. Low brightness projectors in an insufficiently darkened room require a template with good contrast between background, text, and graphics.

Our example presentation will be created with the free *OpenOffice Impress* software, but working with Microsoft PowerPoint and Apple Keynote is very similar. The problem with OpenOffice, of course, is that it is not as stable as its commercial alternatives, it has no support line to call if you have a problem, and its documentation is not as complete as that of Microsoft PowerPoint and Apple Keynote. We launch OpenOffice and choose *Presentation* from the Start Center. The OpenOffice *Presentation Wizard* pops up and asks us to choose between an empty presentation, a presentation from a template, and an existing presentation. We will ignore the templates gallery and design our own simple template, and so we choose *Empty presentation*.

We then select a background for our template. Presentations are typically in landscape orientation although Microsoft PowerPoint also allows the use of portrait orientation, but not of mixed orientations. The *Presentation Wizard* displays a preview of the first slide and a list of background templates. We again ignore the templates and choose *<Original>* from the list and *Screen* as the output medium. The Presentation Wizard then allows us to choose the effects, speeds, and durations of slides, which we ignore. Finally, the software opens two windows side by side; one of these windows shows a presentation in the middle panel of a user interface with a toolbar, slide browser, and task center, while the other window lists the styles and formatting of the slides.

We can now create a new template from *Templates* in the *File* menu and save it as *My First Template* using *Save....* We then select the *Slide Master* option from *Master* in the *View* menu. We choose *Page...* from the *Format* menu, and then select the *Background* tab. A pull-down menu lists *None, Color, Gradient, Hatching,* and *Bitmap,* from which we choose *Color* and then *Chart 1*, which is a classic background for presentations. Since the first release of Microsoft PowerPoint, color gradients from dark blue at the top to lighter blue at the bottom of the slide have been widely used by presenters, but these are now slowly going out of fashion. Alternatives are patterns, textures or (blurred) photos, but these need to be used with care in order to maintain a good contrast between the background and the content of the slide. Hatching is not normally used as a background for presentations. White backgrounds are, however, becoming increasingly popular as they avoid the need for frames around screenshots taken from published graphics with white backgrounds. Some presenters use *Light green* or *Yellow* as their backgrounds, but these are really painful to look at. To sum up the discussion on backgrounds, it is best to use either dark or very light backgrounds, in order to ensure good contrasts with the slides.

We can also use the *Slide Master* to change the default font. Always use sans-serif fonts such as *Helvetica, Arial* or *Gill Sans*; serif fonts contribute to the legibility and readability of printed documents such as newspapers and magazines, but are never used for online documents or presentations as the serifs often appear blurry on computer screens or displays. The master slide shows text fields with all defined styles such as *Title*, *Outline 1* to *9*, and so on. We can click in the *Title* field, mark the title, and then change the font type, size, and color on the master slide. As an example, we change all fonts to Gill Sans, leave the font size as it was, and change the font color to white. Logos and slide numbers can also be added but it is strongly recommended that not too much information be included. We close the Slide Master and save the template as *My First Template* using *Save…* from the *Templates* menu within the *File* menu. Since *My First Template* already exists, the software asks us whether we wish to replace the existing template, and we click on *Yes*.

9.5 Creating Slides

We can now create a new presentation from our sample template. We again choose *Empty Presentation* from the OpenOffice Start Center, choose *From template* from the *Presentation Wizard*, then *My Templates* from the pull-down menu, and finally select *My First Template*, which we have just created. The software creates a new document entitled *Untitled 1* that we can save as *myfirstpresentation_vs1.odp*. The first slide shows the same layout as before, with the *Chart 1* background and *Gill Sans* font.

We first create the title slide (◫Fig. 9.4). The first slide of an empty presentation has two text boxes. We delete the lower text box and move the upper one to the center of the slide. We can do this by right-clicking on the edge of the text box and centering

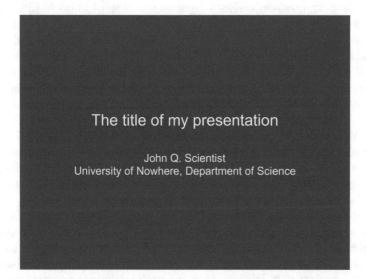

◫ **Fig. 9.4** The first slide of an imaginary presentation. The title of our presentation is *The title of my presentation*, the author is *John Q. Scientist* and his affiliation is with the *University of Nowhere, Department of Science*, as shown in the two lines below the title. We use the *Gill Sans 44 pt light white* font for the title and change the font for the author to *Gill Sans 26 pt light white*, using *Character* from the *Format* menu

the object both vertically and horizontally using *Alignment* from the pull-down menu. The title of our presentation is *The title of my presentation*. The author's name (*John Q. Scientist*) and his affiliation (*University of Nowhere, Department of Science*) appear in the next two lines below the title. We use the *Gill Sans 44 pt light white* font for the title and change the font for the author's details to *Gill Sans 26 pt light white* using *Character* from the *Format* menu. We then save the file as *myfirstpresentation_vs1.odp*.

The next slide is a text slide, although we will generally try to avoid text in our presentation. To start with we again save the document but this time as *myfirstpresentation_vs2.odp*, thus creating version 2 of the document. We create a new slide by right-clicking in the *Slides* panel on the left. Alternatively, we can add a slide using *Slide* from the *Insert* menu. Instead of listing the outline of the talk as discussed in ►Sect. 9.3, we put a hypothesis, a research question, or a scientific controversy on this slide. The *Layouts* panel on the right allows us to choose the *Centered Text* layout. Within the slide the text box asks us to *Click to add text*. We type *What's the research topic?* into the text box and change the font to *Gill Sans 32 pt light white*.

The third slide contains a table, even though tables are not ideal for presentations. If a table is presented then it should be one with very few columns and rows, such as the one edited in ►Sect. 8.5. We again first need to save the document, this time as *myfirstpresentation_vs3.odp*, thus creating version 3 of the document. We then again add a new slide using *Slide* from the *Insert* menu and change the slide layout to *Blank Slide* in the *Slides* panel. The tab-delimited table in ►Sect. 8.5 has seven rows and three columns. We use *Table...* from the *Insert* menu and choose the *Number of columns* as *3* and the *Number of rows* as *7*. After pressing *OK*, a table with a blue fill appears on the slide. Inserting the table from the clipboard (i.e. copied from the file *geochem_data.txt* used in ►Chap. 4), however, creates a separate text object instead of filling the new table fields with the elements from the original table.

We can of course type the numbers into the table fields manually and then edit the layout of the table. After typing in the numbers, we change the font to *Gill Sans 24 pt light black*. The *Table* toolbar that appears after inserting a table can be used to change the design of the table, or to insert and delete single rows and columns, and so on. Having pasted the tab-delimited table into a new text box, we can use *Character...* from the *Format* menu to define *Gill Sans 24 pt light white* as the font. Using *Ruler* from the *View* menu, we can view the rulers and then place left tabs at 4.5 and 9 cm. We can also change the font of the header to *regular* instead of *light*. An empty line can be added after the header, in order to create some space between the header and the table. By right-clicking on the edge of the text box we can center the table, both vertically and horizontally, using *Alignment* from the pull-down menu.

The next slide contains vector graphics, for example the pie chart or the *xyy* plot edited in ►Chap. 8. We again need to first save the document, this time as *myfirstpresentation_vs4.odp*, thus creating version 4 of the document. We again add a new slide using *Slide* from the *Insert* menu. In theory, there are two ways to include vector graphics in a presentation. We can either import them directly as vector graphics, or we can export them from our vector graphics editor as raster graphics and then import the raster graphics into the presentation software. Both methods have their pros and cons. For example, the raster version of a graphic image is essentially opaque and cannot therefore be overlain on other graphic images or text objects in the slide. On the other hand, a high-resolution raster version of a graphic object avoids the distortions in text and graphics that can be caused by software incompatibilities, but this is at the expense of being less editable after inclusion in a slide.

As an example, we import the line graph of the ice core data from R.B. Alley (►Sect. 3.5) that we plotted in ►Sect. 5.2 and edited in ►Sect. 8.2 (◼Fig. 9.5). In the previous chapter we created a version of the line graph with a white font for use in slides with dark backgrounds, saved as *icecore_lineplot_vs7_ai_fordarkbackground.eps*. Using *Open…* from the *File* menu, the plot is opened in *OpenOffice Draw* instead of *OpenOffice Impress*, but we can easily copy and paste the graph into our presentation. Holding down the *Shift* key, we can then proportionally scale the graph to fit the size of the slide. Finally, we can again center the plot by right-clicking on it and then using *Alignment*. We of course have to add a reference to the graph, e.g., *Data from* Alley (2000). Using the *Text* button from the lower toolbar we can place, edit, and move text objects within the slide. We can use *Character…* from the *Format* menu to define *Gill Sans 14 pt light black* as the font. We can also include, scale, and move a pie chart (such as the one in the file *icecore_piechart_vs12_ai_fordarkbackground.eps* that we created for use with dark backgrounds) in our presentation, in a similar way.

The very attractive surface plot, or block diagram, created in ►Sect. 6.3 and subsequently edited in ►Sect. 8.2 is obviously too complex to be integrated into an Open Office presentation (◼Fig. 9.6). It therefore provides us with a good example with which

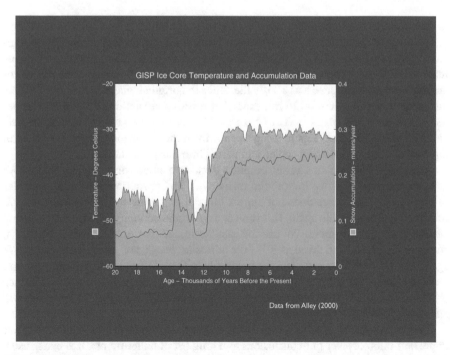

◼ **Fig. 9.5** The fifth slide in our presentation, showing the line graph of the ice core data from R.B. Alley (2000), using data from ►Sect. 3.5 that was plotted in ►Sect. 5.2 and edited in ►Sect. 8.2. We use a version of the line graph with white fonts, for use on slides with dark backgrounds

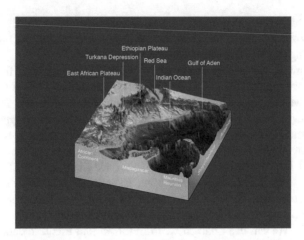

◘ Fig. 9.6 The seventh slide in our presentation, with the very attractive surface plot, or block diagram, created in ►Sect. 6.3 and edited in ►Sect. 8.2. This image provides us with a good example with which to demonstrate the conversion of a complex vector graphics file into a high-resolution raster image, for inclusion in a presentation. The most common format used for presentation slides is the 1,024 × 768 pixel format (the Extended Graphics Array or XGA format—see ►Sect. 9.4 above). We can export the image as a raster image using any vector graphics software, and save it as a TIFF file. Since the original image created in ►Sect. 6.3 had dimensions of 10 cm × 10 cm, exporting it with a resolution of 300 dpi (dots per inch) gives us 1,181 pixels × 1,181 pixels

9

to demonstrate the conversion of a complex vector graphics file into a high-resolution raster image for inclusion in a presentation. The most common format used for presentation slides is the 1,024 × 768 pixel format (the Extended Graphics Array or XGA format—see ►Sect. 9.4 above). We can export the image as a raster image using any vector graphics software and save it as a TIFF file. Since the original image created in ►Sect. 6.3 had dimensions of 10 cm by 10 cm, exporting it with a resolution of 300 dpi (dots per inch) gives us 1,181 pixels by 1,181 pixels. This can then be saved as *etopo2_surfaceplotlight_vs7_ai_block.tif*. Again, using *Open...* from the *File* menu the plot is opened in OpenOffice Draw and can be copied and pasted into our presentation.

The next slide contains raster graphics such as a photo or a satellite image. We need to take care that we import the image with an appropriate resolution in order to ensure that the image does not appear blurred or grainy in the presentation. We first need to save the file again, this time as *myfirstpresentation_vs5.odp*, thus creating version 5 of the document. We use the satellite image that was created in ►Sect. 7.4, edited in ►Sect. 8.3, and saved as JPEG file *naivasha_image_vs4_gimp_smallfile.jpg*. We again create a new slide and open the image file in OpenOffice Draw. We can copy and paste the image from Draw into Impress, scale and center it, and then add a reference (*NASA/GSFC/METI/ERSDAC/JAROS and U.S./Japan ASTER Science Team*) using *Gill Sans 14 pt light white* as the font.

The last slide comes with a take-home message in the form of a summary or an important photo. Never put conclusions as a long list of bullet-points. It is better use a graphic image and to tell people the conclusions, rather than putting them in writing and reading through the list at the end of the presentation. In our example we sim-

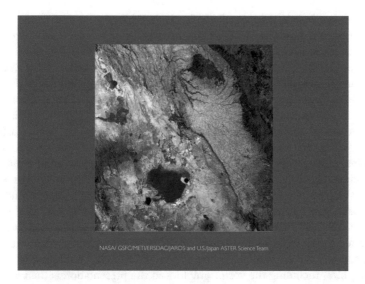

◘ Fig. 9.7 The eighth slide in our presentation, containing a satellite image. We need to take care that we import the image with an appropriate resolution in order to ensure that the image does not appear blurred or grainy in the presentation. We first need to save the file again, this time as *myfirst-presentation_vs5.odp*, thus creating version 5 of the document. We use the satellite image created in ►Sect. 7.4 that was edited in ►Sect. 8.3 and saved as JPEG file *naivasha_image_vs4_gimp_smallfile.jpg* (◘Fig. 9.7). We again create a new slide and open the image file in *OpenOffice Draw*. We can copy and paste the image from Draw into *Impress*, scale and center it, and then add a reference (*NASA/GSFC/ METI/ERSDAC/JAROS and U.S./Japan ASTER Science Team*) using *Gill Sans 14 pt light white* as the font

ply copy the slide with the research question, paste it after the slide with the satellite image, and change the text to our main conclusion. It is not a good idea to then add another slide with acknowledgements, showing a long list of the advisors, colleagues, friends, and sponsors that have all contributed to your work, in the same way that it is not good to include too many names and logos on the first slide of your presentation. You can briefly mention your colleagues and sponsors but it has become a bad tradition to put a lot of logos on the first slide, or even on every slide, of a presentation.

Having finished creating our slides we can now view them all using *Slide Show* from the *Slide Show* menu. However, as must happen quite often at conferences, the presentation may then not look exactly as we expected. It is not possible to predict how the presentation will look on your particular computer, but some typical problems arise on mine when presenting at conferences. For example, all slides sometimes have blue backgrounds, even if some of them have been changed to white. The backgrounds may be white in the preview but in presentation mode they can all be blue. Another classic problem that can arise at conferences when different computers are used for creating slides and for presenting them, even if they both share the same operating system, is that vector graphics may only show as placeholder boxes, displaying the information that the graphics have been created with *Adobe Illustrator*, but no graphics. Even if you try to export the presentation as a PDF file using *Export as PDF…* from the *File* menu, the vector graphics may still not appear in the document, even though once again they are all nicely displayed in the preview mode of the software when editing the slides. The best way to avoid these problems is to export all vector graphics as raster images, merge the images into a single PDF file, and then present the PDF document using the *Full Screen* mode of the PDF software—a strategy that has become very popular when there are presentation incompatibilities at conferences.

Now let us investigate the commercial alternatives to OpenOffice Impress, which are *Microsoft PowerPoint* and *Apple Keynote*. Both products offer tools that are much more professional than those provided by OpenOffice. Having been frustrated by OpenOffice Impress we could simple export the entire presentation into *Microsoft PowerPoint* format under the name *myfirstpresentation_vs6.pptx*, using *Export...* from the *File* menu. Opening the file in PowerPoint reveals some minor changes or incompatibilities between the two software tools. For example, the *light* mode of the *Gill Sans* font has been lost, but we can easily change the font back to *Gill Sans light* within the software. Tables also cause the same problems as with Impress: text therefore needs to be converted into a table with *Microsoft Word* before being included in the presentation. We can remove the two empty boxes in the slides that should have contained the vector graphics. Including the vector graphics in the presentation is then very easy as they can simply be dragged from the file browser and dropped into the slide. Once they are in the presentation they can also be copied into other slides using copy and paste. Scaling and aligning the graphics objects is possible after double-clicking on them. Switching to the presentation mode reveals that this presentation works perfectly well, despite the inclusion of vector graphics.

As an alternative we can open the file *myfirstpresentation_vs6.pptx* with *Apple Keynote*. Drag and drop also works here for importing the vector graphics into the presentation. The positive surprise, however, is that copy and paste also works fine with the table when using Keynote. We simple create an empty table first, modify its layout, and then copy and paste the tab-delimited table from the *geochem_data.txt* text file created in ▶Chap. 4 into the empty table in the slide. We can then change the text font to *Gill Sans 24 pt light black* and change the header text to *regular*. We can export the presentation as a *PDF* file in order to ensure that the document will be compatible with the computer in the conference room. If you want to be absolutely sure that your presentation will run on the conference computer you can also export the slides as raster images (*TIFF*, *JPEG*) and merge the individual images into a single *PDF* document using *Adobe Acrobat*, or any other *PDF* software.

It is, of course, very tempting to add animations to the presentation as most presentation software tools provide large catalogues of animations. Adding video or sound can also make presentations very attractive, for example a video recording of an experiment, or the sound of an ancient musical instrument. There have, however, been numerous bad examples or unfortunate accidents, such as the sound of applause between slides when an author accidentally selected the applause sound transition in PowerPoint but obviously never switched on the speakers of his own computer. The applause, however, was clearly audible after the presentation had been copied onto the laptop provided in the conference hall. Some software provides simple animations that can be very useful in particular instances, such as moving objects around, or changing their scaling. A good way to animate is to assemble an image in several stages, as in the previously discussed examples presented by Steve Jobs and Garr Reynolds, rather than presenting the complete graphic all at once. Another very useful feature allows the inclusion of links to internet webpages that provide access to additional resources, such as audio or video content.

9.6 Practice and Delivery

In ▶Sect. 9.3 we designed the structure of a two-minute presentation. Two minutes is an ideal length for a talk by a beginner as it allows the talk to be practiced repeatedly. If talks are longer, beginners tend to practice the first five minutes quite well but neglect the rest of the talk.

Judith P. Rhodes in her booklet entitled *Scientifically Speaking* (which was first published in 1995 but then updated and revised by others in 2005) gives an excellent summary of, among other things, how to plan, design, and give an oral presentation. A beginner is advised to write down the entire talk from beginning to end, but reading this text at a conference is strictly forbidden. A presenter simply reading a text can be very boring for the audience. Giving an oral presentation is a method of communicating with your colleagues, and it is therefore important to maintain eye contact with the audience while speaking freely, rather than simply focusing on a sheet of paper.

Having written out your talk, you should read it aloud several times, ideally in front of friends or colleagues from your group. Reading aloud helps you to identify any parts of the text that sounds awkward. Moreover, reading the text right through several times helps you to measure the actual time that the talk will require. At conferences, it is regarded as rude and egotistical to run overtime. If the time allotted for your talk is 15 min including questions, you should plan your talk to last for only 12 min. However, be sure to take into account the effect of nervousness, which results in most people speaking more quickly during an actual conference presentation than when practicing.

After reading the presentation aloud several times you should start speaking freely, without referring to your written text. At this point it becomes very obvious if your talk has been well designed, and if the sequence of slides with graphics and photos tell the story nicely without the need for spoken words. In fact, the slides should help you to remember the things that you want to say. A good suggestion that is often made is to make a video recording of your presentation. Watching the video helps you check if you are talking to your laptop screen, rather than maintaining eye contact with the video camera (i.e. your audience). You can also check your body language, such as what you are doing with your hands, and how you are standing. According to Judith P. Rhodes, more than half of the interpersonal communication comes from facial expressions and body language, one third from vocal quality and the tone of voice, and less than ten percent from the content and the actual meaning of the words. In a science talk these relative proportions may well be different, but nonverbal signals clearly remain an important part of the communication process.

A two-minute presentation can easily be practiced 20 or 30 times before having a practice run in front of an expert audience. You should invite your audience to ask questions and to comment on the quality of the presentation. Once you are at the conference you should familiarize yourself with the room that you will use for your presentation. A final practice in that room would be ideal, but is not often possible; you can use a different speaker-ready room instead, if available.

The final delivery should be given using a friendly manner, voice, and body language. Never be aggressive or you will certainly lose your audience; even if they do not leave the room, they are sure to start reading their emails on their laptops using the conference center's Wi-Fi internet. While maintaining eye contact with your audience, try to smile from time to time! This often produces a great response, and individuals might express their agreement with your conclusions by nodding their heads or smiling back at you. By maintaining eye contact you may also notice if the audience has not understood your explanations, which you may then wish to repeat. In this way you will actually communicate with your audience, and your talk will be well delivered.

Recommended Reading

Alley RB (2000) The Younger Dryas cold interval as viewed from central Greenland. Quat Sci Rev 19(1–5):213–226

Duarte N (2009) Slide: ology: the art and science of creating great presentations. O'Reilly Media, Sebastopol, California

Duarte N (2010) Resonate: present visual stories that transform audiences. Wiley, Hoboken, New Jersey

Reynolds G (2010) The naked presenter: delivering powerful presentations with or without slides (voices that matter). New Riders Press, Upper Saddle River, New Jersey

Reynolds G (2011) Presentation Zen: simple ideas on presentation design and delivery (voices that matter), 2nd edn. New Riders Press, Upper Saddle River, New Jersey

Reynolds G (2013) Presentation Zen design: simple design principles and techniques to enhance your presentations, 2nd edn. New Riders Press, Upper Saddle River, New Jersey

Rhodes JP, Gargett A, Abbott M (2005) Scientifically speaking. The Oceanography Society. ► https://tos.org/pdfs/sci_speaking.pdf

Schopf K, Ivany LC, Morris PJ (1992) Onshore-offshore trends in light of ecological locking. GSA abstracts with programs, Cincinnati

9

Creating Conference Posters

Electronic supplementary material The online version of this chapter (https://doi.org/10.1007/978-3-662-56203-1_10) contains supplementary material, which is available to authorized users.

© Springer-Verlag GmbH Germany, part of Springer Nature 2018
M. H. Trauth and E. Sillmann, *Collecting, Processing and Presenting Geoscientific Information*, Springer Textbooks in Earth Sciences, Geography and Environment, https://doi.org/10.1007/978-3-662-56203-1_10

10.1 **Introduction**

The results of a project are typically presented in three formats; as posters, talks, and papers. A poster is collection of figures, photos, and text printed on a large sheet of paper that is presented on a poster-board in a large hall in a conference building. During poster sessions the presenter of the poster can interact personally with the people attending the poster session and visiting the poster. This section is on planning and designing a poster, and also includes suggestions for practicing the presentation of a poster at a conference.

10.2 **Planning a Poster**

Careful planning is essential for a successful poster presentation. Poster sessions at conferences commonly include a large number of poorly designed posters, and a well prepared poster therefore has a good chance of attracting attention. In her booklet entitled *Scientifically Speaking*, Judith P. Rhodes gives an excellent overview of how to plan, design, and present a poster (Rhodes et al. 2005). This booklet is highly recommended and has been used as an inspiration for this section, together with further details added from our own experiences with posters at conferences.

The first objective of a poster is to attract people from a distance of 5–10 m. Poster sessions are typically held in large halls in which conference participants are likely to be walking around, perhaps with a glass of wine, looking at posters in the late afternoon after long sessions of oral-presentations. A large number of posters may be displayed on poster boards within the poster areas, with the authors standing by their posters waiting for visitors and ready to discuss the results of their research. The presenters and their posters are always likely to attract people who are working in the same field and have found the poster in the abstract catalog, but the challenge is to also attract people working in other fields that are nevertheless related to the research topic. These are likely to be scientists that are attending a poster session that covers a broad scientific field, such landslide risks in the Andes, life on Mars, or the causes and consequences of the Younger Dryas cold event.

There are many ways to attract attention to your poster apart from through your own personality, which of course always helps (large groups can often be seen hanging around posters presented by particularly attractive doctoral students). A well-designed poster, however, can also attract visitors even if your work is not exactly in their main field of interest. You then probably have just one or two seconds to prevent a person moving on to the next poster. The two things most likely to help you attract people are a short, effective title and a good overall design. The title should be in a large font, delivering the main message quickly and completely (such as, for example, *There is life on Mars*, or *The oldest Homo ever found*), and a good overall design should incorporate nice colors, attractive graphics and photos, and not too much text.

Below the title you should list the authors, together with their institutional information (but without street addresses), in a smaller font than used for the title. Your email address, as the presenter of the work, should be included with the list of authors. It is a good idea to use first names in the list of authors as this will make it easier for people visiting your poster display to interact informally with you. Some presenters also display their photo on the poster to help visitors locate the author. In her booklet Judith Rhodes suggests omitting institutional logos, but a logo can be useful on a poster if it helps people to identify clusters of posters from the same institution as being related to each other within a larger project. Not many logos should be included, however, and of course not the logos of all of your sponsors.

Having attracted somebody's interest through your great title and design, you need to be able to quickly deliver your main findings and conclusions. If you are at your poster you will probably give a brief oral presentation of your work. In many cases, however, people will spend only two or three minutes at your poster if they are particularly interested, and probably less if your work is only peripheral to their interests. Guiding your visitor through your poster by numbering the figures or using arrows to indicate the sequence helps to keep their attention and encourages them to read the full story. The figure captions should of course always be kept very short. The most common mistake with posters is to include too much information, too many figures, and too much text. Reproducing the original abstract on a poster when it is already included in the conference's abstract volume (or CD) is pointless as most people will not bother to read large amounts of text on a poster anyway. A very brief introduction to the topic, followed by a series of figures with short captions and a brief concluding paragraph, is the maximum amount of text that should be included on a poster. The story should be told with graphics and photos, rather than with text. When assembling the poster, graphics and text should be large enough to be read from 2 to 3 m away. Busy backgrounds such as complex photos should be avoided, and dark colors should be used on light backgrounds rather than the other way round.

The next two sections provide guidelines for creating a poster template from scratch, using *Inkscape* and *Adobe Illustrator* software. We then incorporate graphics and photos edited in Section 8 into the poster. The final section in this chapter then deals with presenting a poster at a conference.

10.3 Creating a Poster Template

A poster generally comprises a background (either white, a single color, or a background photo), a title and list of authors (including their institutional information), an abstract and/or a brief introduction to the topic, a selection of figures and photos with captions, a brief concluding paragraph, and finally, references and acknowledgments, if appropriate. In this section we decide on the background, select the colors to be used throughout the poster, and choose the sans-serif font to be used (e.g., Helvetica, Calibri or Arial) (◻ Fig. 10.1).

We will use Adobe *Illustrator* in this demonstration as it is the most professional software for the demanding task of creating a poster. If Illustrator is not available, however, we can also carry out all the same steps using open source software. Using a landscape format we will present the title, subtitle, author and contact information, a short abstract, references, a table, and five images with captions, but no additional text. We will use a four column layout, boxes with rounded corners, a blue color scheme, and a tinted background. More general information on the software

10

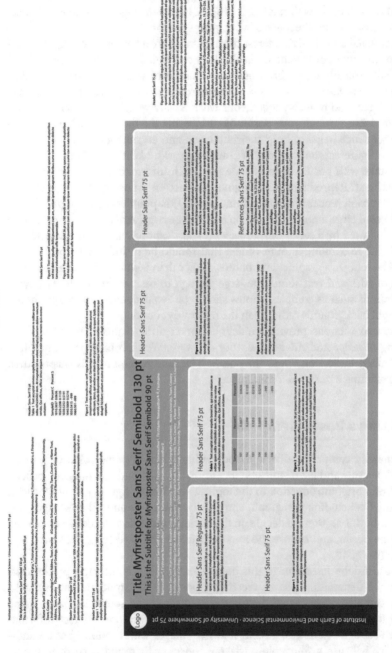

Fig. 10.1 Screenshot of the poster showing poster layout with a provisional distribution of text elements and boxes on a poster size of 197 cm × 100 cm

(including a User's Guide, detailed help, tutorials, and an explanation of technical terms) is available from:

```
https://helpx.adobe.com/illustrator/topics.html
```

Before using the computer we need to draft an outline of the poster on a piece of paper, which can even be as large as the actual poster, in order to plan its design and content. This outline should contain drafts of figures, images, and text, so that the visual structure of the poster can be checked. The design of the poster and the material presented can help to attract browsing visitors, who may not be particularly interested in your topic but could still contribute to your work through discussions from an outsider's perspective. These can sometimes even be more inspiring than the comments that you receive from experts, which may well be comments that you are already familiar with.

A visit to a poster by a conference participant usually follows a more or less typical pattern. Knowing this pattern can be very helpful when creating the poster template. A visitor strolling through the poster session will decide in less than five seconds whether or not a poster looks interesting, so we need to design a template that can capture attention within this short period of time. The poster therefore needs to produce an appealing and cohesive first impression from a distance, which requires a clear structure, great figures, and an esthetic color scheme.

Approaching closer to the poster, a concise title that is intriguing (or even provocative) easy to read (in a large font size), will certainly arouse the curiosity of the visitor. Having first been attracted by the visual effect and the title of the poster, the visitor is likely to be interested in the authors and their affiliations, and then to take some time to actually look at individual graphics and photos. Most people do not want to read very much text and instead start looking around for the author of the poster to talk to. This would be the ideal result, with your poster putting you in touch with a visitor that you might not otherwise have met.

Following these more general considerations, we will now create the template for our poster. We first need to know the size of the conference's poster boards. This information is typically provided in the *Author Guidelines* on the conference webpage. As an example, we can read in the Author Guidelines for the *General Assembly 2016* of the *European Geoscience Union* (EGU) at

```
http://egu2017.eu/guidelines/presenter_guidelines_poster.html
```

the following information:

```
Poster Boards
Poster boards are in landscape format and authors can make use of the full
dimensions of 197 cm width x 100 cm height (e.g. landscape A0 posters as an
often-used format fit perfectly). All the material necessary for attaching
the poster to the poster board is available at the facility desks in the
respective poster area. In addition, there are assistants to help authors
in putting up or in taking down their posters. For each poster board, there
is a desk for placing the private notebook/MacBook for additional
PowerPoint or video presentations. European-type power sockets are
available.
```

The maximum size of the poster is therefore 197 cm × 100 cm. Depending on the size of your printer, the actual size of your poster may well be smaller than this maximum size but must never be smaller. We now launch Illustrator and create a new document named *myfirstposter*, with a *Width* of 1,970 mm and a *Height* of 1,000 mm and with the *bleed* set to 0 mm (◼ Figs. 10.1 and 10.2). We choose the CMYK color mode because we will later want to print it out using an inkjet printer. Additional information on color modes and profiles is provided in ▶ Chap. 11. For the background we draw a rectangle with blue fill, creating a new color with the CMYK code C = 35 M = 20 Y = 10 K = 20 by using the swatches panel and the corresponding tool from the tools panel. For the name and logo of your institution or university we create a rectangular banner on the left side of the poster, 120 mm wide and the same height as the poster, using a blue fill with C = 90 M = 50 Y = 20 K = 30. In order to achieve a uniform arrangement of objects we choose *View > Show Grid*, open the preferences dialog box, choose *Edit > Preferences > Guides & Grid* (in Windows) or *Illustrator > Preferences > Guides & Grid* (in Mac OS), and set the spacing between gridlines using *Gridline Every* 70 mm and *Subdivisions* 14.

Within the blue background we create four boxes in the left half of the poster for figures and text, each measuring 420 mm × 720 mm, and two more boxes in the right half, each measuring 420 mm × 920 mm. We fill all boxes with the blue color C = 10 M = 5 Y = 5 K = 5, which is slightly paler than the blue of the background. We then select all boxes and use *Effect > Stylize > Round Corners* to round the corners of the boxes using a *Radius* of 15 mm, and define a white outline with a width of 20 pt. Having created the boxes we then arrange them provisionally.

We need to leave enough space above the boxes to the left for the title, subtitle, and author information, which we now add. We will then create an auxiliary construction of two rectangles and a system of four *guides* in a new layer we name *guides*. We then create two rectangles in a new layer, for which we use a bright color such as M = 100, with one rectangle measuring 155 × 45 mm in the lower left corner of the poster and the other measuring 35 × 35 mm in the upper right corner. To create guides flanking the rectangles, we position the pointer on the left ruler for a vertical guide and drag the guide into position. We then position the pointer on the top ruler for a horizontal guide and drag the guide into position. As a result, both rectangles are flanked by two horizontal and vertical guides each. Using *View > Guides > Lock Guides, Hide Guides*, or *Clear Guides*, we can edit these guides. The guides help us to align the boxes vertically and horizontally, which we do by first moving the extreme left and the extreme right boxes to their correct positions and then aligning the other boxes between the outer boxes using the *Horizontal Distribute Space* and the *Bottom Alignment* buttons in the *Align* panel. The *Stroke* panel is then used to change the box outlines to white, with a 20 pt *Weight*. Having created and organized various text elements, the boxes, the sidebar, and the background color in the *Layers* panel, we delete the layer *guides* because the auxiliary guidelines are no longer required. Finally, we save the document as the poster template *myfirstposter_1970_1000_empty.ait* and close the document.

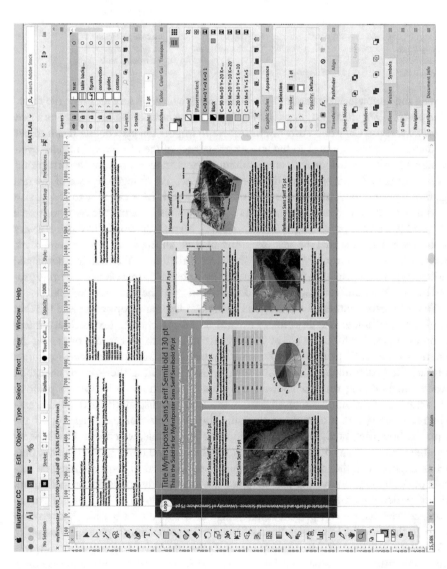

▢ Fig. 10.2 Screenshot illustrating the assembly of a poster using *Adobe Illustrator*, showing the program's menu bar, several panels, horizontal and vertical rulers, guides, and the preparation of text elements on the canvas

10.4   Final Assembly of the Poster

Having created the poster template we now assemble the poster to include the following figures, photos, and placeholder text created in previous sections:

`naivasha_image_vs3_ps_cmyk_2500px.jpg`	*1st column, Figure 1*
`icecore_piechart_vs11_ai_mask.png`	*2nd column, Figure 2*
`icecore_lineplot_vs4_ai_cmyk.eps`	*3rd column, Figure 3*
`etopo2_filledcontourplot_vs1_matlab.eps`	*3rd column, Figure 4*
`etopo1_surfaceplotlight_vs10_ai_flattened.png`	*4th column, Figure 5*
`geochem_data.txt`	*2nd column, table*
`myfirstposter_placeholdertext.rtf`	

Opening the template again creates a new document, which we save as a new file called *myfirstposter_1970_1000_vs1_ai.pdf* (◼ Fig. 10.3). In the layer called *text* we create a new *text frame* and use *Copy and Paste* or *File > Place* to import the placeholder text into our document. The plus sign next to the lower right corner of the *text frame* indicates overflowing text. We can click on it to create additional text frames linked to the first frame, thus providing sufficient space for the overflowing text. We change the style of the text (e.g., font type and font size) to improve its readability and check carefully to ensure that all glyphs have been correctly imported. As stated previously, we use the same sans serif font for all text on the poster (e.g., *Helvetica, Calibri, Myriad,* or *Arial*). Helvetica is a very popular Swiss font created 1956 that is pre-installed on all Apple Macintosh computers, whereas PCs with Microsoft Windows installed provide the Arial alternative developed by Microsoft. The Myriad Pro font comes with the Adobe Creative Cloud. Calibri is provided by MS Office and can also be freely downloaded from the Microsoft website.

We now create another new text frame in the text layer and copy and paste the first line of the placeholder text *Institute of …* into it. All placeholder text includes information on a possible font size (75 pt in our example), which we can use by typing the numerical value of 75 pt into the *Character* panel. Using *Object > Transform > Rotate* we rotate the text frame, including the text, by 90°. Activating the *Preview* checkbox in each panel's dialog window allows the results of all actions to be viewed prior to their actual application. If the text frame is too small to display the text we can easily increase the size of a text object by using the *Selection* tool and dragging a handle on the bounding box. To make the bounding box visible, we choose *View > Show Bounding Box*. We then choose a white color for the text and move it into the dark blue sidebar, aligned with the lower horizontal guide and the vertical center of the blue rectangle of the banner. Above the text we place the logo of our university or institution, which is preferably a white vector graphics object. If there is no suitable vector graphics object available we can place a raster image of the logo over a white background square. In this case the logo resolution will need to be sufficiently high, e.g., 150 dpi.

We next insert the text *Title Myfirstposter … University, Town, Country* into a new text frame, again using a dark blue color for the *Title* and *Subtitle* and the same font sizes as in the placeholder text. The author information should again be in a white font. Footnotes marked by characters such as *a, b, …* or an *asterix* can be used to link the authors to the relevant author information. We then create more text frames for the *Header* with *Sans Serif Regular 75 pt*, for the figure captions (such as *Figure 1 Text …*), for the table captions, and for the list of cited references, to be placed in the various boxes on the poster.

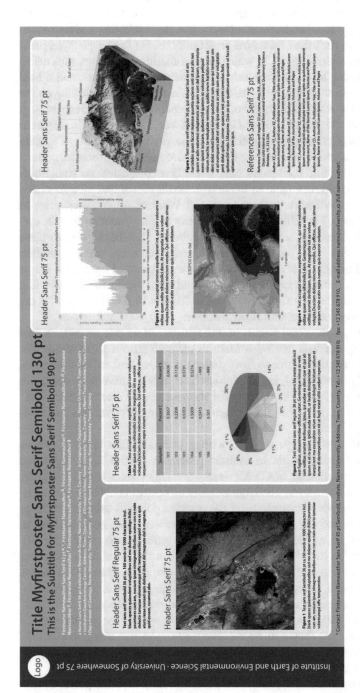

Fig. 10.3 Final version of the poster, with background, sidebar, and boxes in blue shades, placeholder text of ~4,600 letters including blank spaces, five illustrations, and a table. The graphics were generated in previous chapters and edited in ▶ Chap. 8

Using *Window > Type > Paragraph Styles* and *Character Styles* we open two more panels, which allows us to reuse the paragraph and character styles that have been previously defined, allowing a rapid and consistent workflow. After selecting and formatting a text (e.g. a header) we click the *Create New Style* button in the *Paragraph Styles* panel to open the *Paragraph Styles Options*, in which we can modify all settings, thus defining a style. After defining all styles for headers, the main text, references, and a bold *Character* style, these can then be used by selecting a text fragment and clicking on the desired style name in the associated panel. For the references we choose indents of 50 pt for the *Left Indent* and -50 pt for the *First Line Indent*. We then save the document as *myfirstposter_1970_1000_vs1_ai.pdf*, activating the *Preserve Illustrator Editing Capabilities* option and, if we wish, save a second version of the document as a poster template.

We next insert a table into the second box. There is unfortunately no straightforward procedure for formatting a table in Illustrator, which means that all tables need to be edited using other software, such as Microsoft Word or Apple Pages, prior to their inclusion in a poster. The table content is from the *geochem_data.txt* text file created in ▶Chap. 4, which has already been used elsewhere in this book. We can place the edited version of this table, *geochem_data_table.pdf*, which has white separation lines, into the second box in the text layer. For the table background we choose a layer below the table, draw a rectangle using the background color for the table header and a second rectangle using the new CMYK color ($C = 20$ M $= 10$ Y $= 5$ K $= 10$) for the table body. We then save the document as *myfirstposter_1970_1000_vs2_ai.pdf*.

We now insert the graphics and photos into the poster. To do this we create a new layer which we call *figures*, and use *File > Place* to place all figures listed above into this layer, and then move the figures to their desired positions and adjust their sizes. The figure *etopo2_filledcontourplot_vs1_matlab.eps* has not been modified since it was created with MATLAB: it looks like the original shoreline in the *coastline_linegraph_vs1_matlab_17703vertices.eps* file but has a much smaller number of vertices and a relatively small file size (1.3 MB).

MATLAB tends to split long paths when writing postscript files in order to save memory, and has done so in this file. In contrast, generating filled outlines (e.g. when using the `contourf` function in MATLAB—see ▶Sect. 6.3) creates closed polygons, which we position in a lower layer. The large number of vertices (17,703 in our example) may cause problems when trying to print the complex polygons using a PostScript printer, and also when including the figure in presentation slides. As described in ▶Sect. 6.2, we need to reduce the number of vertices in order to overcome this problem, for instance by using the MATLAB script presented in that section. In Illustrator the vertices are called *nodes* and their numbers can also be reduced in that software. As mentioned in ▶Chap. 8, complex vector files may cause printer or software problems, producing error messages such as *Failed to load the requested file*. As an alternative we can export the critical zones as bitmap files and combine the bitmaps with a vector coordinate system, as shown in *etopo2_filledcontourplot_vs2_ai.pdf* and ... *vs3_ai.pdf*. After adjusting the size and position of the figures we save the poster as *myfirstposter_1970_1000_vs3_ai.pdf*.

In order to achieve a uniform appearance we can further modify the poster by first drawing another rectangle, this time 30 mm high, to be used as an auxiliary construction (as above). Using the pointer, this rectangle can be dragged across the workspace like a ruler to measure a constant horizontal distance of 30 mm between headers and figures, and between the white borders of the boxes and the text objects. Wherever we need to, we can drag a guide from the horizontal gauge rule to help us maintain regular distances on the poster. We then save the poster as *myfirstposter_1970_1000_vs5_ai.pdf*, choosing a suitably *high PDF quality* preset, such as *Press Quality* or *PDF/X-1a (2001* and *2003)*, or custom PDF preset files received from our service provider with the file extension. *joboption*, depending on the printer to be used. If the printer prefers to print outlines rather than text we can select all text and choose *Type > Create Outlines*. This is a useful way to avoid printer problems with special fonts, formatting, or glyphs that might not be embedded correctly into a PDF. In this case we save the poster file as version 6, making sure that we also retain the editable version 5 for possible later modifications to the text.

We may also want to create a compact PDF with a reduced file size for digital distribution and therefore save version 7 in RGB color mode, choosing the *Smallest File Size* option from Adobe PDF preset. This allows us to send this file by email so that its contents can be discussed with a supervisor or with colleagues. It also enables us to test the visual effect by asking people from other disciplines for their feedback.

Before printing the final, full sized version of the poster we print some smaller versions to check the contents and layout. We can first print a low-cost, simple version for proofreading, using a laser printer. A second version can then be printed using the actual printer and paper that we will use for the final version but in a smaller size and using less ink, or we can just print out part of the poster by defining an appropriate section using the *Artboard Tool* in Illustrator. Many different types of paper are available for printing posters with an inkjet printer. The standard office paper normally used with desktop printers is of uncoated, matte quality and has a weight of 21–27 lb (80–100 g). Posters are usually printed on heavier paper with a weight of around 55 lb (200 g), either with an uncoated matte surface or with a slightly more expensive coated, glossy or semi-matte surface. For the highest quality printout, coated paper generally has a brighter presentation, absorbs less ink, and yields brilliant colors, similar to the inkjet paper used for printing photos. A coated but semi-matte paper reduces reflections and is therefore best suited for posters that include text, figures, and photos for presentation in a large foyer or hall with complex light sources.

10.5 Presenting a Poster at a Conference

Posters are presented in large foyers or halls, often together with large numbers of other posters, organized according to themes that usually relate to oral sessions presented on the same day. The participants in the oral sessions, including the conveners, the presenters, and the audience, can visit the posters, meet the authors, and discuss with them their results. The conference organizers often offer free wine and soft drinks, making the poster sessions an attractive opportunity to meet other scientists and discuss the overall topic of the poster session. These poster sessions are also important recruiting events and provide many opportunities for young researchers, for example for doctoral students seeking postdoctoral positions for after the completion of their theses. Since the time slots available for oral presentations at large conferences are often very limited, most abstracts submitted with a preference for oral presentations actually end up as posters instead.

The models used to organize a poster session can vary from conference to conference, and even within a single conference, depending on the convener. Some conveners allow poster presenters to give a one-minute presentation of their work before the relevant oral sessions, possibly even including the presentation of a single slide. Others organize guided tours through the poster session, usually consisting of a small group of interested scientists going from poster to poster, with the presenters giving a short oral summary of their work and taking questions from the group. In most cases, however, the conference participants wander around individually and discuss the science with the authors at their posters.

How should one prepare for a poster presentation? Of course presenting a poster to an interested individual or a small group of people is a much less challenging task than giving a talk in a large lecture hall, but nevertheless an inexperienced presenter should write down a few essential facts and conclusions from the poster and prepare a quick summary for presentation when an interested visitor shows up at the poster. A very popular idea is to offer small printouts of the poster (e.g., using A4 format) that people can take home or, if available, reprints of a published journal article that contains the results shown on the poster or related work by the presenter. Some people also like to offer their business cards at their poster boards, or even their CVs if they are looking for a job. Laptops are sometimes placed next to the poster board to display material such as movies, computer animations, project webpages, or software tools.

During the actual poster session, which will typically last one or two hours, the authors of the posters are expected to remain at their poster boards, making it difficult for them to visit other posters in the same session. If there are two or more authors presenting the poster they can obviously take turns at the poster board. If only one author is present, he/she can put a note on the poster informing visitors when they will be available to answer questions.

Recommended Reading

Rhodes JP, Gargett A, Abbott M (2005) Scientifically speaking. The Oceanography Society. ▶https://tos.org/pdfs/sci_speaking.pdf

Creating Manuscripts, Flyers, and Brochures

Electronic supplementary material The online version of this chapter
(https://doi.org/10.1007/978-3-662-56203-1_11) contains supplementary material,
which is available to authorized users.

11.1 **Introduction**

Oral and poster presentations are typically held at conferences or workshops, with a limited number of attendees. Making posters and presentations available online, as either PDF files or videos, increases the potential size of the audience. The most effective way to present scientific information is, however, to publish it as a journal article, thesis, or book. In contrast to these multipage formats, flyers and brochures are used to provide scientific information in a highly condensed form and are typically distributed as handouts at conferences, from trade show booths, during roadshows, or at public presentations.

This chapter deals with the creation of manuscripts, flyers, and brochures as examples of published text documents. We first discuss how to plan and compose a manuscript, and then demonstrate how to create the ready-to-print document. For the first part we use a published research article by John Westgate from the University of Toronto, as a textbook-type example of how to structure a manuscript. The second, more technical part introduces the use of desktop publishing tools to lay out the manuscript. Here we again use graphics that were generated in previous chapters. These can be used in the original vector data formats but are also available in raster file formats (TIFF, PNG, JPEG) from the supplementary material provided online for this book.

11.2 **Planning a Manuscript**

Having completed the experimental part of a project, and then presented and discussed the results at conferences or workshops, scientists are generally required to publish the results of their work as journal articles, theses, project reports and proceedings, or in books. In most cases the manuscripts need to follow certain guidelines, which are usually summarized in a *Guide for Authors* provided on the publisher's webpage. This information, which is available for any journal, helps authors to choose an appropriate journal in which to publish their work. Other criteria such as impact factors are also of great importance in choosing the best possible journal (see ▶ Chap. 2). Multidisciplinary journals such as *Nature* and *Science* only publish relatively short manuscripts. *Science*, for example, only accepts manuscripts of less than 4,500 words (including 40 references) as Research Articles, or of less than 2,500 words as Reports. In most cases the Guide for Authors also contains information on the required structure of a manuscript, which for most journals is as follows:

```
Title
Author names and affiliations
Abstract
Introduction
Regional Setting
Materials and Methods
Results
Discussion
Conclusion
References
```

A research article usually includes tables and figures (including captions), acknowledgements (including information on funding, research permits, and colleagues that have contributed to discussions of the work), appendices, and additional supplementary material.

In ►Chap. 2 we analyzed an article by Neil Roberts published in *Nature* in 1993. We saw how the structure of an article, including its title, the abstract (consisting of the first paragraph), the figures, and the conclusion help the reader to quickly grasp the most important aspects of the research. Since *Nature* and *Science* articles use a slightly different format from that outlined above, let us use the article

```
Westgate JA, Shane PAR, Pearce NJG, Perkins WT, Korisettar R, Chesner
CA, Williams MAJ, Acharyya SK, 1998, All Toba tephra occurrences across
peninsular India belong to the 75,000 yr BP eruption. Quaternary Research,
50, 107–112.
```

This article was published in 1998, in volume 50 of Elsevier's *Quaternary Research* journal (pages 107–112) and is available online at

```
http://www.sciencedirect.com/science/article/pii/S0033589498919743
```

This paper been chosen to explain the structure of a research manuscript because it is a classic example of a journal article with a single conclusion, which is effectively summarized in the 14-word title:

```
All Toba tephra occurrences across peninsular India belong to the 75,000 yr
BP eruption.
```

11

Most readers need just two or three seconds to read and understand the title. Indeed, there is not much to add if you are only interested in the age of the Toba tephra that occur in India. The title, which is very concise, includes the location (*India*), the material (*Toba tephra*) and the age (*all … belong to the 75,000 yr BP eruption*).

The abstract of the article contains additional information on the motivation for the work, the design of the research, the methods employed, the results, the conclusions, and the significance.

A controversy currently exists regarding the number of Toba eruptive events represented in the tephra occurrences across peninsular India. Some claim the presence of a single bed, the 75,000-yr-old Toba tephra; others argue that dating and archaeological evidence suggest the presence of earlier Toba tephra. Resolution of this issue was sought through detailed geochemical analyses of a comprehensive suite of samples, allowing comparison of the Indian samples to those from the Toba caldera in northern Sumatra, Malaysia, and, importantly, the sedimentary core at ODP Site 758 in the Indian Ocean – a core that contains several of the earlier Toba tephra beds. In addition, two samples of Toba tephra from western India were dated by the fission-track method. The results unequivocally demonstrate that all the presently known Toba tephra occurrences in peninsular India belong to the 75,000 yr B.P. Toba eruption. Hence, this tephra bed can be used as an effective tool in the correlation and dating of late Quaternary sedimentary sequences across India and it can no longer be used in support of a middle Pleistocene age for associated Acheulian artifacts.

The structure of the 181-word abstract is as follows. The first sentence, starting with *A controversy currently exists …*, presents an introduction to the topic, and explains why is it important to take the time and effort to analyze the material. The second sentence, which starts with *Some claim* ..., outlines the current hypotheses to be tested. The next two sentences, from *Resolution of this issue* ... to ... *were dated by the fission-track method*, provide information on the nature of the research undertaken and the analytical techniques used. The sentence *The results unequivocally demonstrate* ... then introduces and discusses the results. Finally, the conclusions and implications of the results are presented in the last sentence: *Hence, the tephra* ... *can no longer be used in support of a middle Pleistocene age for associated Acheulian artifacts*. From this analysis the abstract can be seen to be a very short version of the full research article, structured as follows:

```
Introduction (including location)
Materials and methods
Results
Discussion
Conclusion
```

The third level of detail on the research results is then provided by the main text. The *Introduction* is again structured in a way that allows the reader to quickly access the required information. The first few sentences provide an introduction to the topic and some historical background on previous work. The sentence *The number of Toba eruptive events* ... *is presently debated* introduces the scientific problem to be tackled by the research, followed by a summary of the current hypotheses and the significance of the project. The Introduction closes with the sentence *Here, we attempt to answer the question* ..., but most articles use *Here, we present* ... instead, followed by a description of the research, presentation of the results, a discussion, and the conclusions.

In field-based research articles the Introduction is typically followed by sections describing the geographic and geologic settings of the study area, including its topography, rivers and lakes, climate, vegetation, lithological units, and tectonic structures. In the paper by Westgate et al. (1998) the information on the setting is restricted to a location map and a table listing all samples that were analyzed during the project, since the detailed setting is not relevant in this instance. Instead, the sections entitled *Composition of Glass Shards* and *Age Data* contain information on the analytical methods employed, which included the use of an electron microprobe, ICP-MS analyses, and fission-track dating. In this very short paper the results of the chemical analyses and age determinations are also presented within these sections, including three tables and a ternary diagram. The *Age Data* section includes some discussion of the results, concluding that all Toba tephra are from the same eruption. The *Conclusion* summarizes the results with the statement that *New* ... *data show that all the Toba tephra occurrences* ... *belong to the 75,000-yr-old YTT bed* and provides information on the implications and significance of the work. The article then closes with acknowledgements and a list of the references cited.

This analysis of the article by Westgate et al. (1998) should help you to plan your own research article, using a similar structure. Writing a manuscript involves first preparing an outline, often comprising a series of figures that tell a simple story based on a new data set, the results of new research, or a review of existing data. A first draft usually has a maximum length of one or two pages and is structured in the same way as the full manuscript. The draft needs to be discussed with the co-authors and other collaborators in the project before starting to write more detailed text. It is recommend that the first author then creates the full version of the text, for example as *myfirstpaper_vs1.odt* with OpenOffice Writer or as *myfirstpaper_vs1.doc* with Microsoft Word, and sends the file to the co-authors for them to make any changes that they require. The co-authors should then add their initials to the file name (e.g. *myfirstpaper_vs1_mht.odt* or *myfirstpaper_vs1_mht.doc*) before returning the file to the first author. Once all co-authors have agreed on the final version of the manuscript, the first author can submit it to the chosen journal.

When writing the manuscript with a text processor such as OpenOffice Writer, Microsoft Word, or Apple Pages, it is strongly recommended that the character and paragraph formatting feature of the software be used. To demonstrate the use of character and paragraph formatting, let us assume that the manuscript has a classic structure and is composed of the following sections:

```
Title
Author names and affiliations
Abstract
Introduction
Regional Setting
Materials and Methods
Results
Discussion
Conclusion
References
Figure Captions
```

11

Since we have not yet written a scientific text, we use a placeholder text instead. The most popular placeholder text in desktop publishing examples is the free pseudo-latin text *lorem ipsum* (see ▶http://en.wikipedia.org/wiki/Lorem_ipsum for details on the history and discovery of this text):

```
Lorem ipsum: Opera sine nomine scripta

Lorem ipsum dolor sit amet, consectetur adipisici elit, sed eiusmod
tempor incidunt ut labore et dolore magna aliqua. Ut enim ad minim veniam,
quis nostrud exercitation ullamco laboris nisi ut aliquid ex ea commodi
consequat. Quis aute iure reprehenderit in voluptate velit esse cillum
dolore eu fugiat nulla pariatur. Excepteur sint obcaecat cupiditat non
proident, sunt in culpa qui officia deserunt mollit anim id est laborum.
```

The authors' guidelines for particular journals provide all the necessary information on formatting the manuscript file but are often not very detailed with regard to the formatting of the text. We launch *OpenOffice* and choose *Text Document* from the Start Center to start *OpenOffice Writer*. The OpenOffice Writer displays a window with a toolbar and a blank page called *Untitled 1*, with text boundaries and a blinking cursor in the upper left corner, within the text box. As an example, we import the following piece of text from the *myfirstpaper_vs1.txt* file using *Open …* from the *File* menu:

```
This is the title of our manuscript
John Q. Scientist, Jane W. Researcher
```

```
University of Nowhere, Department of Science
Abstract
Lorem ipsum dolor sit amet, consectetur adipisici elit, sed eiusmod tempor
incidunt ut labore et dolore magna aliqua. Ut enim ad minim veniam,
quis nostrud exercitation ullamco laboris nisi ut aliquid ex ea commodi
consequat. Quis aute iure reprehenderit in voluptate velit esse cillum
dolore eu fugiat nulla pariatur. Excepteur sint obcaecat cupiditat non
proident, sunt in culpa qui officia deserunt mollit anim id est laborum.
Introduction
Regional Setting
Materials and Methods
Results
Discussion
Conclusion
References
Figure Captions
```

The software launches a file browser in which we can select the file, and also launches the *ASCII Filter Options* tool. From these options we choose the *Western Europe (ASCII/US)* character set, select *Times New Roman* as the default font, *English (USA)* as the language, and CR as a paragraph break, and then save the document as *myfirst-paper_vs1.odt* (see ▶Sect. 8.4 for more details on character sets, fonts, and paragraph breaks). The default font is *Times New Roman 12 pt*, where Times New Roman is the name of the font and 12 *pt* (or 12 *points*) is the font size. One point is defined as 1/72 inch or 2.54/72 cm, i.e., approximately 0.0139 in. or 0.03528 cm. A larger unit is the *pica* (abbreviated to *p*), which corresponds to 12 pt. Characters in the Times New Roman 12 pt (or 1 p) default font are therefore $12 \times 0.03528 \, \text{cm} = 0.42337 \, \text{cm}$ or about 4.2 mm in height. The Times New Roman font was originally created for the British newspaper *The Times*, in 1931, and its variants are still omnipresent in word processing and book typography.

We may wish to change the *Times New Roman 12 pt* default font to the *Cambria 12 pt* serif font (introduced by Microsoft in 2004), or to another serif font. To do this we use *Select All* from the *Edit* menu to select the entire text, and then choose *Cambria* from the drop-down menu of the *Formatting Toolbar*. We could continue editing the text document by, for example, selecting the headers individually to change their font size relative to the main text, or selecting the names of variables within paragraphs to change them from the default text style to Greek characters. Using the *character* and *paragraph* formatting features of the software, however, makes editing the text significantly easier, especially if the style of a large document has to be modified at a late stage in the writing process, or if the document has to be edited by a graphic designer. Another advantage of this method is that it creates a file (first a Word file, and then later a PDF file) that has all fonts correctly embedded.

Let us now consider the fonts on a computer. A *computer font* is a complete character set of a particular typeface, collected into a *font suitcase* on Apple computers or in the Windows\Fonts folder on PCs. To use a computer font in word processing or desktop publishing software, it must have been installed on our hard drive and made available. Some basic fonts come with a computer's operating system, while other fonts can be delivered and installed with additional software. For example, the now widely used *Cambria* and *Calibri* fonts are included in Microsoft Office and others such as the Microsoft *ClearType* collection are available for free download from the Microsoft webpage; *Minion Pro* and *Myriad Pro* are available with Adobe Typekit, a subscription service for fonts by Adobe Creative Cloud. Additional fonts of reliable quality containing the glyphs necessary for scientific documents can be purchased from commercial type foundries, such as

```
https://typekit.com/
http://www.adobe.com/type/
http://new.myfonts.com/
http://www.linotype.com/
```

After having been installed, a font will remain as a cross-application unit within a computer's operating system even if the software that it came with is uninstalled, and can be accessed through other software tools. Special *font management* software tools help to install, uninstall, and manage fonts on a computer.

Selecting *Styles and Formatting* from the *Format* menu opens a small window listing the default paragraph and character styles, as well as other styles included in the software. Using *Select All* again and double clicking on the *Text body* paragraph style changes the font back to *Times New Roman 12 pt*. We notice that the paragraph style has also changed, in that the space between paragraphs has increased by a few millimeters. Checking *Paragraph...* in the *Format* menu indeed shows a 0.21 cm spacing after paragraphs. We change this spacing to 0.00 cm and the line spacing to *Double*, as required by the Guide for Authors for Quaternary Research. We now again change the font of the *lorem ipsum* paragraph to *Cambria 12 pt*. We then choose *Update Style* using the right button of the *Styles and Formatting* window toolbar. This changes the font for the entire text, even though we only changed the font of the *lorem ipsum* paragraph.

We now continue by changing the font of the headers to *Helvetica 14 pt* and *bold*. We select the *Abstract* header and double click on the *Heading* style, which changes the font to the Microsoft alternative to *Helvetica 14 pt*, which is the *Arial 14 pt* font. We then change the spacing above and below the paragraph to 0.00 cm and the line spacing to *Double* using *Paragraph...* from the *Format* menu, and again update the *Heading* style. We can now select the other headers (*Introduction*, *Regional Setting*, and so forth), and double click on *Heading* to change their style accordingly. Finally, we select the title of the manuscript, change the font to *Helvetica 18 pt* and *bold*, and create a *New Style from Selection* using the right button in the *Styles and Formatting* window toolbar and naming the new style *Title*. To demonstrate how to use character styles, we select the words *labore et dolore* in the text, choose *Variable* from the list of *Character Styles* in the *Styles and Formatting* window toolbar, change the style to *Italics*, and update the style in the same way that we did for the paragraph styles. We have now completed the formatting of the text (◘ Fig. 11.1).

Formatting a large manuscript in this way has the advantage that styles can be subsequently changed in just three simple steps. For example, to change the style of the headings we first select one of the headings, change the font size to 10 pt, and then update the style as we did to update the paragraph styles. We can now see that the font size of all headings has been changed to 10 pt, while the rest of the text remains unchanged. We can modify the *Variable* character style to *underline* instead of *italics* in a similar manner.

11

This is the title of our manuscript

John Q. Scientist, Jane W. Researcher

University of Nowhere, Department of Science

Abstract

Lorem ipsum dolor sit amet, consectetur adipisici elit, sed eiusmod tempor incidunt ut labore et dolore magna aliqua. Ut enim ad minim veniam, quis nostrud exercitation ullamco laboris nisi ut aliquid ex ea commodi consequat. Quis aute iure reprehenderit in voluptate velit esse cillum dolore eu fugiat nulla pariatur. Excepteur sint obcaecat cupiditat non proident, sunt in culpa qui officia deserunt mollit anim id est laborum.

Introduction

Regional Setting

Materials and Methods

Results

Discussion

Conclusion

References

Figure Captions

◻ **Fig. 11.1** Screenshot of the Lorem Ipsum text after editing the layout with OpenOffice Writer using paragraph and character styles. The default font is Cambria 12 pt, the title is in Helvetica 18 pt, and the headers are in Helvetica 14 pt

Formatting manuscripts with a commercial text processor such as Microsoft Word or Apple Pages works in an essentially similar manner to that described above, but these software products are slightly easier to use and, at least in the case of Apple Pages, more stable and reliable than OpenOffice Writer. While working on the above example, OpenOffice Writer froze repeatedly when trying to create a new style. Microsoft Word is usually very slow and often crashes, especially on Macs, and the Microsoft User Folder needs to be cleared from time to time to maintain the performance of the software. In any case, when creating large text documents that include numerous objects other than text (e.g., photos, tables, and graphics) it is advisable to use desktop publishing software such as Scribus or Adobe Indesign, rather than OpenOffice Writer, Microsoft Word, or Apple Pages (see ►Sects. 11.3–11.5).

Manuscripts that are to be submitted to journals, however, do not require any further formatting. Journal articles such as the one discussed previously are never submitted as ready-to-be-used print files in the final layout of the journal. The manuscript file instead comprises a simple text file that includes a cover page, the main text, references, and figure captions. Figures and tables are typically submitted separately as vector or raster files. All files are normally submitted electronically via the journal's web interface. As an example, let us assume the following list of files to be submitted online:

```
johnscientist_maintext.doc
johnscientist_mainfigure_1.eps
johnscientist_mainfigure_2.eps
johnscientist_mainfigure_3.eps
johnscientist_supplementarytext.doc
johnscientist_supplementarytable.doc
johnscientist_supplementaryfigure_1.eps
johnscientist_supplementaryfigure_2.eps
johnscientist_supplementaryfigure_3.eps
```

Most journals do not accept OpenOffice Writer or Apple Pages file formats but both of these software products can export text files in the Microsoft Word format with the .doc file extension. Once the files have been uploaded, the software behind the web interface merges the text, figures, and tables into a single PDF document that then needs to be approved by the submitting author prior to final submission. The web interface also asks for input of information on the authors, including their names, affiliations, phone numbers, and email addresses. The submitting and corresponding author is also required to suggest three to five possible reviewers for the manuscript, excluding colleagues because of possible conflicts of interest.

The merged PDF file of the manuscript then goes to the journal's editorial office where its overall quality and suitability for the specific journal is checked. The editor then contacts two to five reviewers by email and enquires about their availability to review the manuscript. The potential reviewers typically receive the title, the list of authors, and the abstract of the paper to help them decide on their own suitability to evaluate the manuscript. Having agreed to review the manuscript, the reviewers then submit their comments (typically about one page) and a recommendation to accept, to accept with revisions (minor, moderate, or major), or to reject the manuscript. The editor then collects the reviews and decides on the acceptance or rejection of the paper. Once accepted, the manuscript goes to the copy editing office for final layout prior to publication in the journal.

11.3 How to Create Flyers

We will now consider the use of *Desktop Publishing* software to create a *folded flyer*, before later continuing to work on the layout of scientific manuscripts. A folded flyer is usually an attractive, colorful, double-sided product. Flyers are either printed in small numbers (<1,000 copies) on color laser printers and used to introduce research programs at conferences, or they are printed in larger numbers (>1,000 copies) on offset printers and used to promote educational programs at universities, or new commercial products (see ▶ Sect. 11.5 for an introduction to generating PDF files for offset printing).

Whatever its purpose, a flyer should be attractively designed and the information provided easily understood. The text of a flyer needs to be very concise and interesting in order to maintain a lay person's attention. The information provided should be free of complex wording and jargon. The text on a flyer is usually supported by simple graphics and attractive photos, possibly including a simple table summarizing the most important information on, for example, a university curriculum, or a research strategy. The design of flyers (or brochures and leaflets) should make them attractive to the intended audience. They often have backgrounds that are either colored or consist of relevant photos and are therefore commonly printed on glossy coated paper with a grammage of 135–300 g/m^2. They are commonly made from paper of either US Letter or A4 size which, in the case of a flyer, may then be folded into a *C-fold leaflet*. In print production a *flyer* is the technical term for an unfolded piece of paper. For the product we will create in this section, the technical term in pre-press would be *C-folded flyer, leaflet, C-fold leaflet, gate-fold leaflet* or *3-panel roll-fold brochure*, which are all different names for the same product.

Our leaflet will contain the *lorem ipsum* placeholder text from the *myfirstleaflet_ text.docx* document. For the leaflet, we will use the following graphics files that were created in ▶Chaps. 5–7 and subsequently edited in ▶Chap. 8:

```
etopo1_surfaceplotlight_vs5_ps.png
naivasha_image_vs3_ps_cmyk_2500px.jpg
srtm_surfaceplotlight_vs4_ps_cmyk.tif
icecore_lineplot_vs6_ai_blue.tif
```

11.3.1 Desktop Publishing with Adobe InDesign

In this subsection we will design a leaflet using the professional and commercial *Adobe InDesign* software (▶ http://www.adobe.com/indesign.html). After launching the software we set up our workspace using *Application bar > Workspace > Advanced*. If a second display is available we can arrange our favorite panels there in order to keep them within sight and easily accessible while we work on the document using the first screen, and save this as a new workspace. The Adobe InDesign user interface or desktop is very similar in appearance to that of its sister product Adobe Illustrator, with a menu bar, control panel, toolbox, and multiple panels (such as the layers panel, for example) (◻ Fig. 11.2). We choose *Normal Mode* at the bottom of the *toolpanel* to display *non-printing* objects such as object frames, guides, and a white *pasteboard*. On the pasteboard we can provisionally collect and lay out text fragments, images, and illustrations during the design process. Before we start using the software we can hover the pointer over the various tools, read the pop-up *tool tips*, and check the drop-down menus in the application bar. By searching for keywords highlighted in *italic* font within this text, further information, including extensive descriptions of the software and numerous screenshots, can be obtained online from Adobe websites at

```
http://tv.adobe.com/
```

and

```
https://helpx.adobe.com/indesign/user-guide.html
```

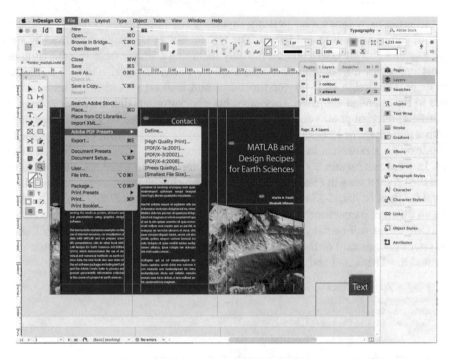

□ **Fig. 11.2**   Screenshot of the *Adobe InDesign* desktop while creating a leaflet, showing the *Document* window with rulers, guides, text frames and the document itself, the *Application* bar, the *Control* panel, the *Toolbar*, several *Panel Groups* in the vertical dock, and the *Layers* panel. Using *File > Adobe PDF Presets* we can select existing presets or define individual presets for generating a PDF

11

We first create a new document using *Application bar > File > New > Document* and obtain the *New Document* window. We then choose two single pages in A4 format and landscape orientation. Using *More Options* we set the margins to 10 mm and the bleeding to 3 mm. Using *Application bar > View > Show Rulers* and *Menu > View > Grids & Guides > Show Guides* we drag vertical guides from the vertical ruler at 97 and 197 mm on the first document page, and at 100 and 200 mm on the second document page. These guides represent the fold lines for the leaflet that separate the *cover page* (Page 1), the four *inner pages* (pages 2–5), and the *back page* (Page 6). We then create three text frames on the first page of the document using *Type Tool* and the *Transform* panel which can be obtained using *Menu > Window > Object & Layout > Transform*. One of these text frames has a width of 77 mm and the other two have widths of 80 mm, and these will become pages 5, 6, and 1 of the leaflet after it has been folded. The text frames on the second document page then have widths of 80, 80, and 77 mm, representing pages 2, 3, and 4 of the folded leaflet. In order to distribute the text frames evenly with a space of 20 mm spacing between them we can use the *Align* panel. After linking the text frames 1–6 in the correct order using their *in port* and *out port*, we want to import the text contained in the *myfirstleaflet_text.docx* Word file, starting with text frame 1 on the right hand side of the first document page (Page 1), which will later form the front cover of the leaflet. Alternatively, we can open the template *myfirstleaflet_template_empty.indt*, which lready contains some ready-to-use formats. The first page of the document shows a red rectangle, together with the word *Text*, both underlain by a *smooth shade*. Using *Application bar > File > File Info* opens the *File Info Dialog* box in which we can add metadata (such as a *Description*, the *Author Information*, and *Keywords*) to our document, which will then remain associated with the document, and even with the PDF file that will be generated later on, for printing.

We now save the file as *myfirstleaflet_vs1.indd* and check the *panels* of our workspace by clicking on their symbols. In the *Layer Panel* we find three layers called *text*, *artwork*, *back color*, with *text* at the top of the list to avoid problems when using transparency, such as *shades* or *filters*. In the *Object Styles* panel we find two default styles marked by square brackets and two new styles called *shadow* and *shadow text*, which we will examine and use later in this subsection. Using *Application bar > View Grids & Guides > Show Guides* we find guides for paper folding and lines indicating text margins. We continue to prepare our workspace and open two more panels using *Application bar > Window > Object & Layout > Transform* and *Align*. Clicking on the first *Sample Page* in the *Pages* panel, we find vertical guides representing widths of 95, 197, 100 and 200 mm, and provisional page numbers 1–6 noted on the pasteboard, beyond the limits of the document so that they will not be printed.

We next explore the two objects that already exist in our document, a red rectangle and a text box. In desktop publishing text objects are tagged with *Paragraph Styles* and *Character Styles*, just as they are in word processing. The expression *Text* in the text box is formatted with a *Paragraph Style* called *Title*. The paragraph style can be defined in the corresponding panel using the *Paragraph Style Options* dialog box. We use the *Myriad* font, a regular style, a 30 pt font size, a 36 pt leading, and the color white. If the chosen font is not available, InDesign highlights the text in red so that we can replace it. The *Drop Shadow* of the text object is changed by using an object style called *shadow text* with an opacity of 90%, a 1 mm distance, and a $-130°$ angle, which can be defined using the *Object Style* panel. The red rectangle lies in the *artwork* layer and its drop shadow is defined by the *shadow* object style. We choose the *Selection* tool from the toolbox, select the text and rectangle objects, and either delete them or move them to somewhere outside the document on the pasteboard.

Opening the *Swatches* panel shows four default colors *Registration*, *None*, *Paper*, and *Black,* as well as two shades of blue identified by their CMYK codes. Please note that the color *Registration* is reserved for correlating and aligning the overlapping CMYK colors of the four printing plates. This particular color has the CMYK values of $C = 100\ M = 100\ Y = 100\ K = 100$, i.e. all colors are at maximum intensities or 100%. The registration marks appear on each of the four printing plates, e.g., as a crosshair target outside the actual page. While four times full intensity (equal to 400%) is acceptable for the registration marks, it is actually too much printing color on the paper in the offset printing process, which has an upper limited of 300–320%. On a normal page, black should therefore generally consist only of $C = 0\ M = 0\ Y = 0\ K = 100$. Another common mistake in offset printing is the use of *RGB-black*, which consists of full-intensity red, green and blue. The printer converts the three RGB colors into

CMYK colors red, green and blue, without including key, and black line art, text, and tables therefore appear with blurred outlines or as typefaces with overlapping colors.

We can now create a new color, a dark greenish blue, for the background of the flyer (or leaflet), which we do in the *Swatches* panel using *Swatches Application bar > New Color Swatch*. We type in the values $C = 90 M = 20 Y = 25 K = 70$ for the CMYK *Process Color* and click *OK* to add the new color to the panel. We then open the *Pages* panel, select the *back color* layer, and draw a 300 mm × 216 mm rectangle in the new color. The size of the rectangle also includes a circumferential *bleed*, which is an extra 2–3 mm beyond the edge of the 297 mm × 210 mm sheet to avoid any white, unprinted edges in the final, trimmed document.

We next use the *Type* tool from the toolbox to create a text frame in the *text* layer, in between the guides on the cover page. Using *File > Place …* we select the placeholder text *myfirstleaflet_text.docx* and place it on the right hand side of Page 1 of the leaflet. When importing the text we can choose between various options, such as the way in which the styles of the original document are preserved during import. In our example we decide to retain all the formatting from the word processing software. The original paragraph styles and character styles are automatically converted into the InDesign styles *Title, Header2 18 pt, Header1 24 pt, Normal, text 10 pt folder*, and *Author 10 pt bold*. We delete the black RGB swatch in the *Swatches* panel and replace it with *Paper*, which automatically changes the black color to white, for all styles.

Continuing with the *Type* tool, we create five new text frames on pages 2–6 of the leaflet. We connect these new text frames with each other by clicking on the *out port* of one box (indicated by a + symbol in the lower right corner of the box), and on the *in port* in the upper left corner of the subsequent box. The text flowing into multiple text boxes can be inspected using *Application bar > Choose View > Show Text Threads*. Within the flowing text, we can add frame breaks using *Application bar > Type > Insert Break Character > Frame Break* to divide the text into several segments, each comprising a header and a short text.

Having created the background of the flyer and *placed* the text, we will now include four graphics in the *artwork* layer. The four graphics (listed above) were created with MATLAB in ▶Chaps. 5–7 and subsequently edited in ▶Chap. 8 for inclusion in presentations, posters, and manuscripts. We use the *Rectangle Frame Tool* to create a rectangular frame on each of the first four pages of the leaflet (pages 1, 2, 3, and 4) and save the file as *myfirstleaflet_vs2.indd*. Using *Application bar > File > Place*, we import each of the four figures and scale them to fit within the frames using *Application bar > Object > Fitting > Fill Frame Proportionally*. We then use the *Selection* and *Hand* tools with the computer mouse, or alternatively the *Direct Selection* tool, to select and then move or scale each graphic object within its frame, without actually modifying the frame itself. To adjust the arrangement and size of the illustrations we can use the *Transform* and *Align* panels. If we wish, we can add drop shadows from the *Object Styles* panel: we can choose between *shadow* object style for illustrations and *shadow text* for text objects such as the title on the cover page.

In our example we have always used *Application bar > File > Place* to place objects into the InDesign document, instead of including them in the document. The advantage of placing objects rather than including them in the document is that each of these objects can then be further edited using Illustrator or Photoshop, and updated automatically within the InDesign document. Furthermore, when placing instead of including graphics objects the file size of the InDesign document remains relatively small, even if working with high-resolution images. While working on the document in InDesign, the software displays a low screen-resolution version of the image, which facilitates the modification of larger documents.

Working with placed objects requires full control over the various files contributing to the main document. The *Links* panel is used as a tool for managing these files, for navigating to the locations of the various objects included in the document, for opening the *placed* objects (such as PDFs, graphics, and images) with their original editing software, and for updating file paths and versions. The *Links* panel can also be used simply as an information center, providing all details concerning the *placed* objects, the software used for their creation, their date of creation, any use of *transparency*, the location of the objects within the document, their *paths* (i.e. their locations within the computer's file system), and their resolution.

The illustration *icecore_lineplot_vs6_ai_fordarkbackground.eps* has been specifically designed to stand out well on the petrol blue background, as it has white labels. The graph, however, would look even better on the dark background if the line color was also white and if the area between the two curves was light blue. We can open the illustration in its original software (Illustrator) by first clicking on the figure's name and then on the *Pencil* icon at the bottom right of the *Links* panel. In Illustrator we can then change all the line colors to white, remembering to deactivate the overprinting checkbox in the *Attributes* panel.

We next remove the graph's white background. To do this we open the *Pathfinder* panel using *Application bar > Window > Pathfinder* while the two curves are still selected, press the *Exclude Overlapping Shape Areas* button, and fill the area between the curves with a new color that has the CMYK code $C = 90\,M = 20\,Y = 25\,K = 70$ and a 40% color intensity. After editing the line graph we save it under the name *icecore_lineplot_vs6_ai_blue_neg.eps* and update the *Links* panel in the main document in InDesign. By using the *content grabber* we can scale and move the line graph within the graphics frame in order to highlight, or zoom into, a detail of the illustration, provided it is used for decorative purposes only. The content grabber is a particularly useful tool when working on images within the document.

A very nice feature of text frames is the ability to float text around an image using the *Wrap* function of the software. Using *Application bar > Window > Text Wrap*, or alternatively, by creating an object style for wrapping, we can choose *Wrap Around Object Shape* from the various options available. Another elegant way to wrap text around graphics objects is to manipulate the shape of the text frame using the *Add Anchor Point Tool*. This process is very similar to editing a path in vector graphics software, in that we can add or remove anchor points and then move them around to change the shape of a polygon. After completing the layout of the flyer with InDesign we again save the document as *myfirstleaflet_vs2.indd*.

11.3.2 Preparing the Document for Printing Using Adobe InDesign

Having finished our work on the layout of the leaflet (or brochure and leaflet) we need to check the document before generating a PDF file. We first examine the list of fonts used in the document, using the *Find Font* dialog box in *Type > Find Font*. In our example document we have used three different fonts: *Minion Regular*, *Myriad Regular*, and *Myriad Semibold*. In order to change a font, or to search and replace for a missing font, we choose *Replace With*, select another typeface, and then click *Find First* to check where the font has been used. We then replace it using *Change All*, with the checkbox *Redefine Style When Changing All* activated. Both paragraph and character styles can be changed with this command, throughout the whole document. Having checked the fonts we then check the resolution of the file in the *Link* panel. The resolution for a digital printout should be in the range of 150 to 300 dpi. Having checked both the fonts and the resolution of the document, we save it as *myfirstleaflet_vs3.indd*.

To collect and store the main document, together with all linked files, into a single folder, we use the *Package* feature of InDesign. *Application bar > File > Package…* creates a new folder with the same name as our main document. Within this folder all files related to our leaflet project are collected into a single location in our file system. The *Package* feature also creates a subdirectory called *links* inside this new folder, gathering together all of the linked files contributing to the leaflet. Having completed the packaging we can finally generate a PDF file using *Application bar > File > Adobe PDF Presets > High Quality Print*, or any other appropriate choice of a *Print Preset*. When generating the PDF document with color or images running up to edge of the format, we activate the *bleed* and *registration (trim) marks* options, which are positioned outside the A4 paper format. To make sure that these objects will appear on the paper print out we use an A3 paper format to print it at full size (100%) and later cut off the *bleed*. Whereas the *High Quality Print* settings are sufficient for printing the document on a digital laser printer, generating PDF documents for offset printing requires additional settings to be made, such as for *color management*, *preflight*, and specific *export options* (see ▶Sect. 11.5 for more details on offset printing).

11.4 Designing a Thesis or a Research Report

The writing and formatting of a multi-page doctoral thesis, a comprehensive research report, or any other long text document, requires more detailed planning than a relatively short, 10–15 page journal article. Although there is an increasing tendency in most countries to submit cumulative doctoral theses consisting of several (commonly three) thematically related journal articles, most doctoral candidates submit their theses in edited formats, with graphics objects, tables, and photos included in the page layout. In most cases, text processors such as OpenOffice Writer, Microsoft Word, or Apple Pages can easily manage this task as long as the text documents do not become too large and contain only a few other elements besides the text. This section aims to help us decide on the best tool to use for editing a thesis or research report, i.e. a text document that includes several chapters and sections, with a large amount of text and numerous graphics, tables, images, and appendices.

A popular way to handle large documents using a text processor such as Writer, Word, or Pages is to split them into smaller, more manageable files representing individual chapters. Having finished writing the text we need to switch to desktop publishing software such as the commercial *Adobe InDesign*. The fundamental difference between word processing software and desktop publishing (DTP) software is a result of their different origins. The first word processors were line editors designed for typewriter-style terminals and they therefore had very limited page-making capabilities. In contrast, desktop publishing tools evolved from typesetting printed pages by arranging rectangular objects of text and illustrations within a printing frame.

The distinction between these two ways of creating printed documents is, however, becoming increasingly blurred. Word processors now provide some of the typical desktop publishing features that are included in Scribus and InDesign. For example, Pages by Apple now has both word processing and page layout features, as you can create text boxes that float freely around graphics and images. Of course these features are very limited compared with those provided by InDesign, but the software easily outperforms Word on both Macs and PCs in its handling, stability, and reliability.

For the user, there are three main differences between text processors and DTP software. The first difference is that, as has been stated previously, both text and images are independent objects in DTP software, organized vertically into *layers* and horizontally within these layers into *frames*. This allows the design of complex page layouts in a workflow that is fundamentally different from that of text processors. In fact, editing page layouts with DTP software has many similarities with vector graphics editors such as Adobe Illustrator. The second difference is that DTP uses an elaborate option for inserting or *importing* objects such as images, graphics, or PDF files (and many other types of files) by *placing* in the document rather than embedding them, as has already been demonstrated in ▶Sect. 11.3 for flyers (or brochures and leaflets). Placing objects instead of importing or embedding them reduces the file size and can therefore dramatically improve the performance of the software, especially when working with large manuscripts that include a large number of illustrations. As described in ▶Sect. 11.3, the *path* of the original file stored somewhere in the file system appears in the *Links* panel. This panel is useful for supervising, managing, and updating large numbers of placed graphics objects. The third difference between text processors and DTP software is that text processors provide a choice between *CMYK* and *RGB color modes*, which is essential when preparing a manuscript for offset printing, whereas DTP software does not provide this choice but applies color modes consistently throughout the document.

In addition to these three main differences, professional DTP applications offer numerous features that can increase productivity, reliability, and flexibility when creating large documents with complex structures and ambitious layouts. For example, the *layers*, *master pages*, *object styles*, *text styles*, and *indexing* tools, together with the ability to *combine* and *synchronize* multiple documents using the *book function*, are of great assistance to a publisher. Advanced features are available for the creation of publication-quality documents, such as the *preflight* and *production* tools, the handling of *fonts*, the ability to add *bookmarks*, the possibility of using *multiple page sizes*, and support for *PDF/X standards* when generating PDFs. The InDesign software also provides a seamless interface with other members of the Adobe Creative Cloud suite, such as Photoshop, Illustrator, Acrobat, and Bridge.

In this section we use Adobe InDesign to create a typesetting file for a thesis or a research report. Once again, the reader is advised to make use of the comprehensive help and tutorials provided on the software vendor's webpage:

```
https://helpx.adobe.com/indesign/user-guide.html
```

As in the previous subsection on desktop publishing with Adobe InDesign (in ►Sect. 11.3), keywords *highlighted* in *italics* in this section refer to the corresponding Adobe help resources available online.

Authors typically write a thesis or research report on a word processor, rather than using a desktop publishing tool. In such cases it is advisable to use the specific paragraph and character styles recommended in the publisher's *Manuscript Guidelines* or *Author Instructions* throughout the document as this will greatly facilitate further processing with InDesign. We have already learned (in ►Sect. 11.3) how to create a small leaflet with this software. We again use the placeholder text *lorem ipsum* as an example text for our longer document; this is provided in the Word document called *myfirst-thesis_chapter_1_placeholder_text.docx* and uses paragraph and character styles. The Word file does not contain any graphics as these are provided in a separate folder for typesetting. The document instead lists all figures and table captions at the end of the text. The list refers to graphs and images edited in ►Chap. 8, which we will also use in our thesis or research report. The following tutorial uses InDesign to create a document with a single-sided layout for digital printing. Later on (in ►Sect. 11.5) we will use the same material but change the manuscript into a book with double-sided pages for professional printing.

11

11.4.1 Creating a New Document with Layers, Master Pages, and Pagination

After launching InDesign, *Application bar > File > New > Document* opens the *New Document* window where we create eight A4 pages with a portrait orientation. To define the *type area* we choose *More Options* and set the margins in accordance with publisher's guidelines. In our example we set the top and left margins to 25 mm, the bottom margin to 15 mm, and the right margin to 20 mm. Since most theses are printed using digital laser printers we do not need to include bleed areas in the document as we did for the leaflet in ►Sect. 11.3. We then create four layers called *text, artwork, figure caption,* and *running head,* from top to bottom. Finally, we add metadata such as the author information, the title of the document, and so forth, to the file using *Application bar > File > File Info.* If required, we can set up a suitable workspace with *Application bar > Workspace > Advanced,* or alternatively we can use the two screen workspace that we set up in the previous section.

Having completed the setup of our DTP document we then proceed to work on the individual pages. The *Pages* panel displays eight pages and a single master page. Selecting the *Master* page A opens the corresponding master page, on which we can work as on a regular page, but all modifications and additions will then feature on every page of our document to which this particular master page is applied. Using the text tool we create a thin text frame for the *running head* element, immediately above the print area in the corresponding layer, and type a placeholder text such as *The Title of my First Thesis*. For this short piece of text we create and apply a new *Paragraph* style. We call it *running head*, open its *Paragraph* style in the *Paragraph* style panel, define the *Basic Character Formats* as *Font Family* Myriad, and set the *Size* to 10 pt, the *Leading* to *Automatic*, and the *Case* to *All Caps*. Changing the *Case* to *All Caps* turns all characters into capital letters. To create a dividing line we use *Paragraph Rules*, click the *Rule Below* checkbox, set the *Weight* to 0.75 pt, and choose a black solid line of the same *Width* as the column. We set the *Language* to English USA, the *Indents and Spacing* to 15 mm *Left Indent*, and the *Alignment* to centre.

For automatic pagination of the document we create a small text frame in the upper right corner of the master page, choose *Application bar > Type > Insert Special Character > Markers > Current Page Number*, and insert a text variable for the page number. We duplicate the *running head* style, change the font style to *Semibold*, deactivate the *Rules* checkbox, set all indents to 0 mm, set the *Alignment* to left, name the new style *pagina*, and align the running head and pagina with each other. Scrolling through the normal pages in the *Pages* panel or document window shows that all pages now have a running head and a continuous pagination.

11.4.2 Importing Text, Using Text Frames, and Using Flowing Text

Having created a new document with layers, master pages, and pagination, we will now place text into the text frames. Using the *Type* tool we create a text frame on the first page, flush with the page margins. If necessary, we can use the *Selection* tool to adjust the size of the frame and move it, using the frame handles. We then activate the text frame and use *Application bar > Place …* to select *myfirstthesis_chapter_1_placeholder_text.docx*, which contains the text to be placed in the document. In the *Show Import Options* dialog window we choose to import the text, together with the text format. Using *Open*, we import the text into our typesetting file. We find all paragraph styles marked with a small floppy disk icon, indicating that all new styles need to be checked; we will return to this topic in the next subsection. All styles used by our Word template are now included in our typesetting file.

Having imported the text we notice that only the first few paragraphs are actually visible because the text is larger than the text frames on the first page. This *overset text* requires the creation of additional text frames, linked to the first text frame. We click on the red plus-symbol inside the *Out port* in the lower right corner of the frame and then use the *Selection* tool to show the *loaded text icon*. We can drag out a second text frame on the second page and again fit it to the page margins. Alternatively, we can use the

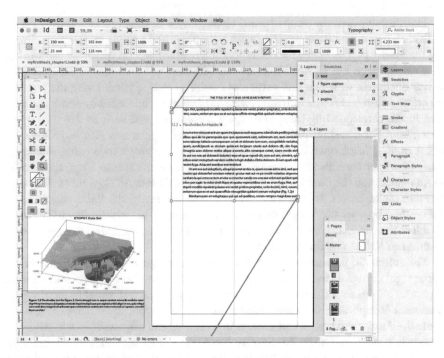

◙ Fig. 11.3 Screenshot of the *Adobe InDesign* desktop while creating a thesis, showing a figure and its caption provisionally prepared on the pasteboard. The text frame on the document page is already reduced to a smaller height where space for the figure is needed, letting the threaded text flow between the interconnected text frames

11

Autoflow feature of the software to create multiple text frames on many pages by holding down the *Shift* key while creating the next frame on page three. Using *Application bar > Choose View > Extras > Show Text Threads* we can visualize the links between the text frames. The *Pages* panel now lists eight document pages, of which seven contain text. We save a version of our file as *myfirstthesis_chapter1_vs1.indd* (◙Fig. 11.3).

When importing text we try to keep all the original fonts and styles of the Word document. Using *Application bar > Type > Find Font* we examine the list of fonts for the document: a yellow triangle indicates any fonts that are not available on our computer and therefore need to be replaced. Activating the *Redefine Style When Changing All* checkbox and choosing *Change All* updates all paragraph and character styles, as well as eliminating any font conflicts within the document—an important feature to prevent serious problems during printing. Redefining the styles actually ensures that all fonts will be embedded into the later PDF file and avoids unpleasant surprises when having the PDF printed in a commercial print shop. This contrasts with the clandestine replacement of fonts by word processing software, instead alerting the user if there are missing fonts.

Alternatively, we can use the *Find Font* feature to replace fonts throughout the document without having to modify each individual paragraph and character style manually. This is very useful if, for example, we wish to switch from one variant of the

Times font family to another (such as from *Times* to *Times New Roman*), or if we wish to make a more general change from a *serif* font to a *sans-serif* font. We can also use this feature to alter the appearance of text fragments from different parts of a cumulative thesis, or of the contributions made by different members of a research project to the final report. In our example we change all fonts into a sans-serif typeface such as *Myriad*, with the *Italic, Regular,* and *Bold* style types. All style definitions and all text in our document are then adjusted simultaneously.

11.4.3 Direct Formatting, Paragraph Styles, and Character Styles

Each InDesign document has a *Basic* paragraph and character style that is applied to unformatted typed or pasted text. We can define and modify paragraph and character styles in the same way as with word processors, choosing the font family, style type, size, color, alignment and other attributes, in the *Paragraph* and *Character* panels or in the *Control* panel. An easier and more thorough way to define automated formats is to use InDesign's centralized attributes called *Character Styles* and *Paragraph Styles*. Clicking into the text and then on the *Create New Style* icon at the bottom of the panel opens the *New Paragraph Style* window in which the formatting is already compiled as the *Style Settings* and where we can give a name to the style. To the left we find a list of more options for fine adjustment of the document's format. To apply the text format to a new paragraph we simply click into a text paragraph and then choose the desired style from the paragraph style panel.

Let us now take a closer look at the existing styles in our document, which we imported together with the text. By selecting the red text on the first page, the associated *normal* paragraph style and the *comment in red* character style are highlighted with light blue in the corresponding panels. Clicking the highlighted character style opens the *Character Styles Options* dialog box. The *Basic Character Format* is Myriad Pro Regular and the *Character Color* has no stroke but a fill color. As is typical with word processing software, however, the fill color is in the RGB mode, which is indicated by the three-colored rectangle to the right of the color called *RTF r190 g0 b0*. We change this color into a CMYK color with $C = 15\ M = 100\ Y = 100\ K = 0$. We notice that only a small number of attributes differ from the *normal* paragraph style and therefore delete the Word style and replace it with the InDesign *normal* paragraph style. The style called *figure caption bold* in the panel still shows the floppy disk symbol, but it disappears after clicking the style and checking the settings in the dialog window. The original font of the figure and table captions at the end of the document is changed to *Myriad Pro Bold*, according to the definition of the corresponding *Paragraph* style.

We will next edit the style for bullet point lists such as the one after the red *Here comes a list of four bullet points* comment in the placeholder text. Clicking into the list highlights the *normal bullets* in the *Paragraph Styles* panel, indicating the style that is already applied to the list, even though the list does not yet have any bullet points. Opening the *Options* dialogue box, we switch the *List Type* from *None* to *Bullets* in the

Bullets and Numbering selection and choose a dot as the marker from the *Bullet Character* display. Should the *Bullet Character* display still be empty, we choose *Add* to open a new dialog box and then select the *Myriad Pro* font and a symbol suitable for use in scientific texts from the glyph list. Using *Text After*, we insert a *Tab* and set the text alignment to left, the indent to 15 mm, and the tabulator position to 21 mm. In the *Indent and Spacing* options we set the *First line indent* to 10 mm. To check the results we click on the *Preview* checkbox in the lower left corner of the dialog box so that we can toggle between viewing the old or the new definitions.

In addition to these basic modifications to the document we can introduce automated numbering of the paragraph styles as *heading 1*, *heading 2*, and *heading 3*. We can do this by choosing *Numbers* as the *List Type* in the *Bullets and Numbering* options, starting a new list, and setting the *Mode* to *Start at 1* for the highest level, i.e. *heading 1*. We can check whether our automated numbering works correctly since the previous numbering is still present as plain text. We save our file as *myfirstthesis_chapter1.indd* and also save an additional copy of it as *myfirstthesis_chapter1_vs1.indd*, overwriting the previous (preliminary) version of the file.

11.4.4 Creating Book Files, Page Numbers, and a Table of Contents

In this subsection we consider examples of what are known as the *Long document features* of desktop publishing tools. The first example of these features that we will explore is the InDesign *Book* function, which helps to make large publications manageable. We first create a new empty document called *frontmatter* containing all settings used in the thesis or research report. We open *myfirstthesis_chapter2.indd* by clicking the file on the *Book* panel, save it as *myfirstthesis_frontmatter.indd*, and remove all text, figures, and captions. We import all text contained in the Word file *myfirstthesis_frontmatter_placeholder_cover.docx*, place it to Page I, and and then import all text in the Word file *…_abstract.docx* and place it in a separate text frame on Page IV. We then duplicate and modify the existing paragraph styles to format the cover page and save the file as *myfirstthesis_frontmatter.indd*.

We next use *Application bar > File > New > Book* to create a new file and save it as *myfirstthesis_book.indb*. The *Book* panel appears and we can add new chapters by clicking on the *plus* icon in the lower right corner of the panel. We then add three existing documents from the supplementary electronic material provided online for this book to the new frontmatter and ▶Chap. 1 documents:

```
myfirstthesis_chapter2.indd
myfirstthesis_chapter3.indd
myfirstthesis_backmatter.indd
```

In the book panel's *Document Numbering Options* context menu we choose Roman figures *I, II, III, IV …* starting with the first page of the frontmatter, to keep the pagination independent of the main content of the thesis. For the first chapter of the book we again start with page number 1 but now use Arabic figures *1, 2, 3, 4….* For the sub-

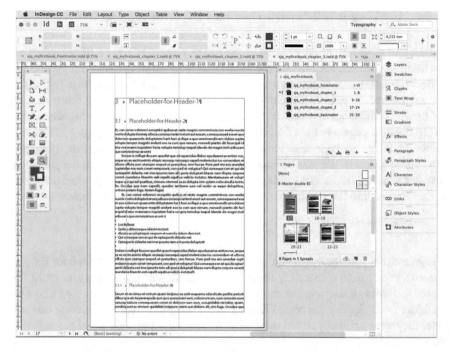

☐ Fig. 11.4 Screenshot of the *Adobe InDesign* desktop while creating a thesis, showing the first page of chapter 3, with text formatted using three levels for the headers and body text with the formats *normal*, *indent* and *normal bullets*. The running head on the first page and pagination of a new chapter have already been removed. The *Book* panel lists all parts of the book, with *Automatic Page Numbering* activated and Roman page numbers for the frontmatter and Arabic page numbers for the other pages of the book. A red or green bullet at the foot of the document window indicates the result of the *preflight* for this chapter

sequent chapters and the backmatter we again select Arabic figures, but then choose *Automatic Page Numbering*, continuing from the previous book document (☐ Fig. 11.4).

We then open the frontmatter document and adjust the contents settings using *Application bar > Layout > Table of Contents* to create a preliminary version of the *Table of Contents* (TOC). In the right column, we click on « *Add* to add the *header 1, header 2,* and *header 3* styles to the *Include Paragraph Styles* list on the left. We then click the button *More Options*, select the *Include Book Documents* checkbox, and use the *content 1, content 2,* and *content 3* paragraph styles. We adjust the paragraph style options for each entry and, in addition, use the *dots content* character style to format the table of contents automatically. After generating the TOC a text cursor pops up that enables us to create an independent text frame flowing onto pages II and III. Checking all entries we find a deliberate spelling mistake in the placeholder text *Palceholder with Typo for Header 2*. In order to fix this error we do not edit it in the TOC but instead open the first chapter, correct the typo there and then update the TOC. In the next subsection we place images and a table into ▶Chap. 1 and then again update the TOC.

11.4.5   Placing Images and Tables, and Adding Captions

Figures and tables are usually placed at the top or bottom of a page, after having been discussed in the text. We can use the *Find/Change* function to search for *fig* and *table* to find any references to figures and tables within the text. We find that *Figure 1.1* is first mentioned on page 2 of the document. We then click on the text frame in which the figure is discussed and reduce the height of the frame, thus moving some of the text to the next page. We then find that *Figure 1.2* is mentioned on page 3 of the document and *Table 1 on* page 6, so we also modify the sizes of the frames on those pages in a similar manner. We provisionally place the following graphics on the pasteboard outside the actual document (as described in ▶Sect. 11.3), in a layer called *artwork*, to be later moved into the blank spaces that we have created within the document:

```
icecore_lineplot_vs4_ai_cmyk.jpg
etopo1_surfaceplotcontours_vs3_ps.jpg
```

We again either *place* or *import* the graphics, as we did in ▶Sect. 11.3 (and for the same reasons), rather than embedding them into the document using *copy and paste*. This approach allows all files used in the document to be subsequently collected into a single folder using *Application bar > Choose File > Package*, no matter where on our computer they were originally stored originally. We again use the *Links* panel to manage the placed graphics.

When working on a manuscript we need to have access to the names and *Status* of each of the graphics, and to additional information such as the *Page* on which they appear, *ICC Profile*, *Effective PPI* and *Transparency*. The status of graphics is shown as *Up to Date*, *Modified*, *Missing*, or *Embedded*. If the status needs to be changed, the *Links* panel is quite helpful for finding, relinking, and editing graphics. When updating the status we can also adjust the preview resolution for the graphics as they are displayed on our computer screen using *View > Display Performance > High Quality Display* or … > *Fast Display*, depending on our preference. As stated previously, the figure and table captions are normally included at the end of the document. We can copy them from there and paste them into individual new text boxes in the *figure caption* layer. We can then give them the *figure caption* style and, if necessary, resize them to fit into the frame.

Using *Application bar > File > Place* we import the *geochem_data.txt* file that was edited as tab-delimited text in ▶Sect. 8.5. Alternatively, we can just copy and paste the content from a word processor or spreadsheet software file. Using *Application bar > Table > Convert Text to Table* we convert the tab-delimited table into a formatted table by choosing *Tab* as the column separator and *Paragraph* as the row separator. We expand the size of the table and then create the *table head* paragraph style with *12 pt bold* characters and the alignment set to *Centre*. We use the same *12 pt regular* font as in the main document, set the alignment to *Right*, and set a *Right Indent* of *17 mm*. Using *Application bar > Table > Table Options* and *Cell Options* we give the table border and strokes a *1 pt* thickness and set the border and strokes color to *Paper*. We set the fill color of the body cells to *Blue 15% Tint* and the fill color of the header row to

Blue 30% Tint. As was the case in ►Chap. 4, the absurd value of −999 indicates missing data in the table. We replace the short hyphen for the minus sign with the special longer hyphen known as *En-Dash* using *Application bar > Type > Insert Special Character > Hyphens and Dashes > En-Dash*. We then save the first chapter of the thesis or research report as *myfirstthesis_chapter1.indd*. We must remember to update our TOC using *Application bar > Layout > Update Table of Contents* in the frontmatter document, which is listed in the *Book* panel. We then save this file as *myfirstthesis_front-matter.indd*.

11.4.6 Preflight, Packaging, and Generating a PDF File

The *Book* panel is an efficient control center when completing large documents, being used to check all technical aspects of the files and to generate the PDF file for printing. The panel can be used to *synchronize* or *manage book* documents, to *update* their numbering, and to *preflight*, *package*, and *export* the elements of the book.

Before passing on the files for printing it is essential to perform a quality check on all of our files. This so-called *Preflight* simulates the printing process by searching the entire book for inconsistencies such as missing fonts, status problems with linked illustrations, low-resolution images, overset text, and so forth. Whenever a red bullet appears next to a file in the *Book* panel it indicates that an error has been detected, whereas a green bullet indicates no errors in the book file.

The *Preflight* feature generates a report on the results of the search. We can launch the *Preflight* panel using *Window > Output > Preflight*. After fixing all existing problems (as described in the subsection entitled *Preparing the document for printing with Adobe InDesign* in ►Sect. 11.3) we use *File > Package*, check once again the summary of all items included in the book, and then save everything in new folder which we name *11_4_myfirstthesis_book_folder*. As mentioned in ►Sect. 11.3, packaging creates a directory inside this folder called *links*, in which all linked graphics files are stored. This is very useful if numerous files that were previously spread all over the file system on your computer are now collected together into the *links* folder, as this makes it easier to hand over the book files to another person, or to generate an updated PDF file after modifying the typesetting file.

The final step in this subsection is to create an Adobe PDF file, either using *Application bar > File > Export* or by creating it directly from the *Book* panel using *Adobe PDF Presets > High Quality Print* (or any other appropriate *Print Preset* for digital or desktop printing). In addition to the print file we can create a second, smaller version of the file, suitable for publishing online or sending by email. We can accomplish this by using a different, very convenient *Adobe PDF Preset* such as *Adobe PDF (Interactive)*, which includes interactive elements, or the *Smallest File Size* feature, which reduces the file size by reducing the resolution of all images. In the professional Adobe *Acrobat* PDF software we can examine our PDF file, fill out the *Description* form, set the *Initial View* of the document in the *Page layout* to *Single Page*, and set the *Magnification* to *Fit Page*. If many copies of the thesis or research report will be required then offset printing is preferable to digital laser printing. In ▶Sect. 11.5 we will demonstrate how to lay out a double-sided product, again using InDesign's *Book File* feature. The more technical aspects of printing a large numbers of copies, such as the use of *color management* and the relevance of specific *export options* for generating *PDF/X-1a standard*, high resolution, press-ready PDFs, are also described in ▶Sect. 11.5.

11.5 Assembling and Laying Out Books

Printed books are generally paperback or hardcover multi-page text documents, typically with more than two hundred pages, either in black and white or including colored graphics and images, sometimes with CDs or DVDs attached. The readership of a science book may range from undergraduate students looking for an introduction to a specific topic or wanting to learn about a new method using a tutorial-style textbook (such as the one you are currently reading), to experts seeking information on the results of a major research program.

Since the first commercial e-reader was introduced in 1997 the *electronic book* or *e-book* has developed from a niche product to a widespread technology, paralleled by the rapid development of computers, personal digital assistants (PDAs), smartphones, electronic book readers, and other electronic devices such as desktop and tablet computers, on which e-books can be purchased, stored, and displayed. The boundaries between classic publication formats, such as those used for journals or books, and online media used to present scientific content, are consequently becoming increasingly blurred. In such publications, graphics and audio-visual elements that one usually expects to be included on Internet websites (such as video clips, interactive elements, and hyperlinks) can be embedded into a fixed layout format *PDF* or a floating layout format *ePUB* (short for *electronic publication*). Both text processors and DTP software offer built-in converters to create such ePUB files. Printed editions may try to compensate for this added value by providing supplementary data on a CD, on a DVD, or online.

Before either an e-book or a traditional book can be held in your hand it will need to be assembled for publication. Such assembly typically starts with the meticulous work of collecting and creating content, drafting an outline, writing the main text, compiling tables, listing references, and designing figures. In order to avoid unpleasant surprises when offset printing the final product or publishing an ePUB document online, most authors submit raw text and graphics material rather than providing the publisher with a print-ready file. The publisher's webpage usually provides the authors with guidelines for organizing the material into an appropriate format, for example in *Book Manuscript Guidelines*, or *Author Instructions*, or even a *Microsoft Word document template*. Springer, for example, provides concise information at

```
http://www.springer.com/authors/book+authors?SGWID=0-154102-0-0-0
```

as well as providing a ready-to-use Word template and LaTeX macro packages at http://www.springer.com/authors/book+authors?SGWID=0-154102-12-417900-0

We can download the *Word template* file from the Springer webpage and use it for laying out new text. After activating the *TextTools* and *TitlePage* toolbars using *View > Toolbars* we can either type new text into the template or copy and paste a text fragment from another text editor into the file. The template contains various predefined *paragraph styles*, together with *character styles* such as *H1, H2, H3* for headers, typographic alignment styles such as *Justified* and *Indent*, and the *Index Entry* tool for including keywords in the book's index. As already stated in ▶Sect. 8.4, strict use of *paragraph styles* and *character styles* from the very beginning when writing a book simplifies the layout process significantly, including the compilation of a table of contents and an index.

High quality figures are essential in all earth science publications. Perfectly reproduced illustrations attract attention to the publication and its author, as well as adding refinement to a scientific work. A publisher's *Artwork and Illustrations Guidelines* provide concise technical information for preparing and submitting figures electronically.

11.5.1 Color and Color Management

As demonstrated in ▶Chap. 8, the choice of an appropriate *Color Management* such as, for example, the definition of an appropriate color profile for the publication, is essential if a high quality product is to be achieved. General information on color profiles and detailed descriptions of available color profiles are available at

http://www.color.org/

for US color standards and

http://www.eci.org/

for European standards. Adobe also provides its own information resources, for example at

https://helpx.adobe.com/indesign/user-guide.html

where we can choose *Color* to access all relevant information on working with colors when using Adobe software products. A fundamental requirement with colors is to distinguish between the *RGB color mode* used for computers, projectors and digital cameras, and the *CMYK color mode* used for four-color (*4C*) printing. For example, most digital photographs include a standard red, green, blue (*sRGB*) color profile. The color modes and profiles of photographs should not be modified during processing unless the final output format (printed book, e-book, or online publication) and its specific color management settings have already been defined. For example, an incorrect choice of a CMYK color mode may result in large files with pale colors when published online, while RGB photographs typically appear dark and grayish when printed in a book.

11.5.2 Compilation of the Manuscript Files and Submission to the Publisher

Authors usually submit their book manuscripts to a publisher either on a CD or DVD, or online via the publisher's FTP server. Manuscripts are usually first submitted as a single PDF file that includes all text, figures, and tables. Once the submission has been accepted the manuscript is delivered to the publisher as an editable open-format file, for example as an RTF or a Word document. The PDF version of the manuscript needs to be carefully checked for the correct appearance of all special characters and fonts, in particular where they occur within figures. In most programs *File > Print* or *File > Export* opens the print dialog box in which we can choose *Adobe PDF* as the export format and then click *Save* in order to transform the manuscript into a PDF, which the author can then approve prior to submission. When submitting multiple files rather than a single file into which all files have been merged, the file names should be carefully chosen in order to allow the editor to easily associate each file with the relevant book chapter, figure, attachment, and so forth. Below is an example of a hierarchical structure used for the compilation of files to be submitted to a publisher:

```
miller_booktitle_year
        miller_chapter_1
                readme_ miller1.doc
                miller1.doc
                miller1.pdf
                miller1_fig_1_1.tiff
                miller1_fig_1_2.eps
                miller1_fig_1_3.eps
                miller1_fig_1_4.eps
        miller_chapter_2
                readme_ miller2.doc
                miller2.doc
                miller2.pdf
                miller2_fig_captions.doc
                miller2_fig_2_1.eps
                miller2_fig_2_2.tiff
                miller2_fig_2_3.eps
                miller2_fig_2_4.tiff
                miller2_fig_2_5.eps
        miller_chapter_3
                miller2.doc
                (cont'd)
```

In this example *miller_booktitle_year* is the name of the folder containing the entire book. The folders *miller_chapter_1* to *miller_chapter_3* contain all files comprising chapters 1–3, while the optional *readme* files provide supplementary information on the contents of the folders and the *.doc* files are the text files for the chapters (including the figure captions). The *.pdf* files are preliminary PDF versions of the chapters (including text, tables and figures). All other files (such as the *.eps* files and *.tiff* files) are the figures, with the relevant chapter number included in the file name. After the publisher's production department has completed the typesetting some publishers allow authors to read and edit their articles, book chapters, or entire book proofs online.

11

11.5.3 Working on a Book Project with InDesign

This subsection considers some of the advanced features of InDesign that are particularly useful when creating books for offset printing, such as the *color management* and *preflight* features, and some specific *export options*. Please note that we use *Chapter* and *Section* (with upper case *C* and *S*) to refer to chapters in the book that you are reading, and *chapter* (with lower case *c*) when referring to chapters of the theoretical book being discussed in this demonstration. We will use the text and figures in *myfirstthesis* from ▶Sect. 11.4 to create a double page spread book. We first create six new folders for the frontmatter, backmatter, and the four chapters of the thesis. Each of the four folders of the thesis chapters contains a *links* folder, which we copy and paste into the equivalent folder of the new book.

```
sjq_myfirstbook_a_frontmatter folder
sjq_myfirstbook_chapter_1 folder
sjq_myfirstbook_chapter_2 folder
sjq_myfirstbook_chapter_3 folder
sjq_myfirstbook_x_backmatter folder
```

Within each *links* folder we add the author's initials *sjq* (for Scientist, John Q.) and a descriptive or *aptronym* part to the figures' file names. Below is a list that compares new file names used in this section with the old file names (in brackets) that were defined in ▶Chap. 8 and then used in ▶Sect. 11.4, starting with those in the *links* folder for chapter 1 of the new book

```
sjq_myfirstbook_fig_1_1_icecore.eps
(icecore_lineplot_vs5_ai_cmyk.eps)

sjq_myfirstbook_fig_1_2_surfaceplotcontours.jpg
(etopo1_surfaceplotcontours_vs3_ps.jpg)
```

followed by those in the links folders for chapter 2

```
sjq_myfirstbook_fig_2_1_filledcontourplot.eps
(coastline_linegraph_vs3_matlab_7009vertices.eps)

sjq_myfirstbook_fig_2_2_srtmsurfaceplotlight.jpg
(srtm_surfaceplotlight_vs4_ps_cmyk.tif)

sjq_myfirstbook_fig_2_3_coastlinelinegraph.eps
(coastline_linegraph_vs3_matlab_7009vertices.eps)

sjq_myfirstbook_fig_2_4_srtmfaultssatmerged.eps
(srtm_faults_sat_merged_vs4_ai.jpg)
```

and chapter 3,

```
sjq_myfirstbook_fig_3_1_etopo1surfaceplotlight
(etopo1_surfaceplotlight_vs10_ai_flattened.png)

sjq_myfirstbook_fig_3_2_rosediagram.eps
(directional_rosediagram_vs3_ai_fill1.png)
```

```
sjq_myfirstbook_fig_3_3_etopolpseudocolorplot.eps
(etopol_pseudocolorplot_vs3_ai.eps)
```

We then open a new empty InDesign document using *Application bar > File > New Document* and set up four *Facing Pages* with a B5 *Page Size*, which has a *Width* of 176 mm and a *Height* of 250 mm. The margins of the page are set to 25 mm for the *Top*, 15 mm for the *Bottom*, 25 mm for the *Inside*, and 20 mm for the *Outside*, with a *Bleed* of 3 mm. We can now import the text contained in *myfirstthesis_chapter_1_ placeholder_text.docx*, as described in ▶Sect. 11.4. Activating *Import Styles Automatically* in the *Import Options* ensures that the text keeps its formatting (such as its paragraph and character styles). For the other components of the book (such as the frontmatter, chapters 2 and 3, and the backmatter) we can use draft InDesign documents with file names that end with *..._draft_vs1.indd*. We can import all styles from one of these draft documents into our new document, and then create a new *Book* file called *sjq_myfirstbook_draft.indb* in which to compile and manage each chapter individually. The book uses relatively small font sizes compared to the thesis, such as 20 pt, 16 pt, and 12 pt for headers 1, 2, and 3, 10 pt for the main text, and 9 pt for the figure captions.

For chapter 1 we can also use an InDesign template, such as *sjq_myfirstbook_ empty.indt*, which creates a new empty document called *Untitled-1*. The original template with the file extension *.indt* remains unchanged, and we now either place and import the placeholder text, figures and captions from the *links* folder into the new document. We need to search any missing fonts and replace them with fonts available on our computer by choosing *Redefine Style When Changing all* and *Change All*. Opening the *Pages* panel shows a double paged document, starting with a single right-hand page (page 1), three double pages (left- and right-handed pages 2+3, 4+5, and 6+7), and ending with a left-hand page (page 8).

We open the *Master* page by clicking the small preview icon in the *Pages* panel. In the *Master* page we can add and modify items such as running heads, pagination, text frames, and the file modification date in the center of the double page, all of which are located in separate layers. The text modification date in the *DRAFT* layer, for instance, can be altered by using *Choose Type > Text Variables > Define* and then *Application bar > Type > Text Variables > Insert Variables > Modification Date*. The file modification date is then automatically updated when opening and saving the document. We can use another *Text Variable* to create *dynamic column titles*; this generates column titles and automatically updates them whenever the document is modified. In order to achieve this we define a *Text Variable* by selecting the last *header 1* on a page, and change the settings of this variable to *running head* on the right hand side in the *Master* page. We save our file as *sjq_myfirstbook_chapter_1_draft_vs1.indd*; this file contains only text at this stage of our work but we can gradually place figures and captions into it and save the intermediate steps as different versions labeled from *..._draft_vs2* to *..._draft_vs5*.

Another advanced feature when using InDesign to create books is the *Baseline Grids* feature. Baseline grids help us when adjusting the text to line up the columns on a single page, on two facing pages, or between the two sides of the (slightly transparent) pages of the book. We can explore the default baseline grid in our document using *View > Grids & Guides > Show Baseline Grid*. We can click within an arbitrary text paragraph and then on the *Align To Baseline Grid* button in the *Control* panel. To adjust the text to the grid for the entire document, we open the *Paragraph Style* panel, click on the *normal* style, select *Indents and Spacing* in its *Paragraph Style Options*, and activate the *Align to Grid All Lines* option.

The two other styles used for the main text that may require adjustment are the *normal indent* and *normal bullet* styles. Using *Edit > Preferences > Grids* (for Windows) or *InDesign > Preferences > Grids* (for Mac OS) we keep the default *Light Blue* color for the grid, let it start at 78.5 pt *Relative to Top of Page*, and select *Increment Every* 12 pt (because the body text of our document has a 12 point leading, as defined in the *Basic Character Formats* of the *Paragraph Style Options*). In order to achieve a suitable adjustment of the headers 1, 2, and 3 we specify their *Leading, Space Before,* and *Space After* values such that the sum of all these values is equal to 12 pt, or a multiple of, 12 pt. As an example, we use Leading = 12 pt, Space Before = 24 pt, and Space After = 12 pt for header 1, values of 18 pt, 18 pt, and 12 pt for header 2, and values of 16 pt, 12 pt, and 8 pt for header 3. Having completed our work on the baseline grid we save the document as *sjq_myfirstbook_chapter_1_grid.indd*.

The last advanced editing feature of InDesign that we will consider (which is also available with most word processing tools) is the ability to use non-breaking or hard spaces to prevent unwanted line breaks (for instance between numbers and units, or between *Fig.* and the figure). We can use *Type > Insert White Space > Nonbreaking Space (Fixed Width)* to insert a non-breaking space. With *Fig. 1* and other figure numbers, for example, which may occur many times within a text, we can also use *Edit > Find/Change*, type *Fig.* followed by a blank space into the *Find What* box, and then replace this string by *Fig.* followed by a *White Space > Nonbreaking Space* with a constant width.

Having completed the drafts of the chapters we can create the new book file *sjq_myfirstbook_draft.indb* in which to compile and manage all files. We then place our figures and captions, format the table, and create a table of contents (TOC) for the frontmatter, as described in ▶Sect. 11.4. We save copies of the book chapters and define names ending with, for example,..._draft_vs2.indd. We then generate a PDF file called *sjq_myfirstbook_draft.pdf* using *Export Book to PDF* from the *Book* panel menu, choosing the *Smallest File Size* option in order to create a small file size suitable for emailing to the proofreader.

A proofreader's job is often considered to involve simply correcting grammatical and typographic errors. However, a good proofreader with a scientific background that is appropriate to your discipline will do far more than that. Many of the alterations that such a proofreader makes may not in fact be *correction* at all, but suggested improvements to the wording that will make the text read more easily and sound more natural. This will (hopefully) not only make your paper a pleasure to read but also help to ensure that the meaning is always clear. It is important, however, to always bear in mind that a proofreader does not necessarily understand exactly what you are trying to say or what message you are trying to get across. All of the alterations that he or she makes must therefore be treated as being only suggestions. You, as the author, are always responsible for ensuring that the end result still says exactly what you want it to say: you can either reject or accept the proofreader's suggestions.

If a proofreader is not sure what you are trying to say then he or she will generally query it in a comment on the side of the page, often providing one or more suggested alternatives to the original version. Proofreaders usually review and comment the PDF file using Adobe Acrobat.

```
http://adobe.com/acrobat
https://helpx.adobe.com/acrobat/user-guide.html
```

After launching the software we open *sjq_myfirstbook_draft.pdf*. We can use *View > Page Display > Two-Up* to change the page display to *Show Cover Page During Two-Up*, and browse the document pages. Using *File > Properties > Description* we fill in the *Metadata*, which contains relevant information about the document, such as its *title*, *author*, and some *keywords*. Using *Choose View > Toolbars > Comment & Markup* we can launch the *Commenting and Markup Tools* to add *Sticky Notes*, to *Cross Out* or *Highlight* text, or to draw geometric shapes such as lines or arrows within the document. In the event that our proofreader (or colleague) does not have access to the professional version of Acrobat, the free *Adobe Reader* software can also be used for reviewing and commenting PDF files if we have activated *Comments > Enable For Commenting In Adobe Reader* in Acrobat before saving the file. After receiving a commented PDF file we can click on *Show Comments List* in the *Comment and Markup* toolbar to display all comments and a toolbar for *Managing Comments*, which includes functions such as *sorting*, *filtering*, and *replying to comments*. The author or designer can then modify the original InDesign file and create a new PDF file.

11.5.4 Generating the Final PDF File for Printing

After the proofreading, we save a fresh set of folders and files for the document using *File > Package* (this time without affixing the word..._draft*) which we call *sjq_myfirstbook.indb, sjq_myfirstbook_chapter_1.indd*, and so on. We clear up all book documents, remove all elements from the *pasteboard* and from within the document (such as in the *DRAFT* layer) as they are no longer needed, and check that all the fonts and figures used in the book are available. We can then run a *preflight* for the whole book in which a printing process is simulated by the software and may yield a number of error messages indicating different kinds of problems with the file (such as missing fonts, for example), but ideally responds with green bullets in the panel indicating a successful preflight with no problems encountered.

We then check all colors in the file. Opening the *Swatches* panel of chapter 1, we find four default color symbols listed: *None* (for no color), *Paper, Black*, and *Registration*, the latter only being used for technical purposes such as for crop marks in the document. The blue color $C = 80, M = 45, Y = 20, K = 15$ is used for headers and the table background, and $K = 30$ (or 30%) black is used for the file modification date in the center of the double page. Importantly, we notice that there is a single red color that resulted from the text import from the Word file. This is an RGB color, as indicated by the three-colored square to the right of the panel, and therefore not suitable for offset printing. We therefore select this color, which is called *WORD_R190_G0_80*, and all other superfluous colors in the panel, click the *Delete* icon at the bottom of the *Swatches* panel, and use *Remove and Replace* to remove the red color and replace it with the default *Black*.

Using *Window > Output > Separations Preview* produces a list with five rows in an independently floating panel. The first row shows a combination of the four CMYK *printing plates*, while the other rows each show only a single plate for *Cyan*, *Magenta*, *Yellow*, or *Black* (*Key*). We wish to use only *process colors* in our document, these being combinations of *Cyan*, *Magenta*, *Yellow*, and *Black* (*Key*). All other colors are converted into the four process colors using the *Swatch* panel. When toggling the *visibility* of the black plate, black text, or outlines, we notice that the main text, the strokes, illustration labels, and all other black graphics objects such as black lines in graphs, disappear. In this context, the black color *overprinting* feature that was demonstrated in ▶Chap. 8 with the *icecore_piechart_vs12_ai_overprint.eps* file is of great importance when printing the document on an offset printer. In Illustrator we activate the overprinting feature for black text objects by changing the *Attributes* checkbox to *Overprint Fill*. As indicated previously in ▶Chap. 8, the overprinting feature needs to be treated with care and only applied to black; overprinting of lighter colors, in particular white, results in the complete invisibility of all objects in these colors. Furthermore, overprinting is only possible for vector graphics and not for raster images.

We then list all fonts used in the document by using *Type > Find Font* and eliminating those that are not used in the document or not available on our computer, as these can not be embedded into the PDF file. We use the *Find Next* tool to find the *Wingdings Regular* font, which highlights the list of bullets on the first page of the document. Opening the associated *Paragraph Styles* panel we find that the font is used for the *normal bullets* style, and opening *Bullets and Numbering* in the *Paragraph Styles Options* we find the glyph 131, a black square from *Wingdings Regular* used as a *Bullet Character*.

After completing all these tasks we run the *Preflight Book* tool from the *Book* panel to check all documents included in the book. Again, when all errors listed in the preflight report have been fixed, green bullets appear on the right side of the panel. We can now select *Export Book to PDF* from the panel menu, select the *Adobe PDF (Print)* option, and generate the PDF file, which we then save as *sjq_myfirstbook.pdf*. When we generate the PDF file we use an appropriate *Adobe PDF Preset*, having first consulted the printer on this matter. The *PDF/X-1a:2001* format, as an example of an *Adobe PDF Preset*, is a very strict standard for offset printing that does not allow any transparency, embedded interactive features, or the RGB color mode. Large book projects can also be exported chapter by chapter using *File > Export* instead of generating a single PDF file for the entire book, in order to avoid large files. If the printing equipment requires PDF files with crop and bleed marks we can activate the *Crop Marks* or *Bleed Marks* checkboxes in the *Export Adobe PDF* dialog box, as well as selecting the correct *Bleed*.

Having created the PDF file we open it in Acrobat, adjust the page view, and check the *Output Preview* (which is very similar to the *Separations Preview* in InDesign) using *Advanced > Print Production > Show Print Production Toolbar*. Note that in ▶Chap. 8 we did not set the overprinting feature for black in the figure called *sjq_* *myfirstbook_fig_2_1_filledcontourplot.eps*, and we therefore have to modify this in Illustrator and update it in InDesign prior to actually generating the PDF file. For the final preflight with Acrobat we use the *Sheetfed offset (CMYK)* profile (or any other suitable profile), which yields a report listing information about the file, including any warnings. These details concerning the file need to be discussed with the printer, who will also run a preflight and use PDF modification software and fixing routines (such as *Solvero* or *PitStop Pro*) to improve the document and ensure that it perfectly matches the printing process.

Recommended Reading

Westgate JA, Shane PAR, Pearce NJG, Perkins WT, Korisettar R, Chesner CA, Williams MAJ, Acharyya SK (1998) All Toba tephra occurrences across peninsular India belong to the 75,000 yr BP eruption. Quat Res 50:107–112

Creating Multimedia Publications

Electronic supplementary material The online version of this chapter
(https://doi.org/10.1007/978-3-662-56203-1_12) contains supplementary material,
which is available to authorized users.

12.1 **Introduction**

Many scientific publications now include multimedia material, either as an electronic supplement or embedded within publication files; chapter deals mainly with such embedded material. Multimedia objects can be integrated into web pages since a very long time. The *HyperText Markup Language* (HTML), developed in 1990 by Tim Berners-Lee at CERN and now available as HTML5, the fifth version of the HTML standard, was designed to allow authors to provide not only text, images and interactive forms, but also audio and video, within a single (web)page.

Multimedia presentations are also very popular at conferences, e.g. presentations that include audio or video material. Since 1997 it has been possible to include such audio and video material, as well as other interactive options such as web integration, in Microsoft PowerPoint presentations, as is also the case with Apple Keynote (first released in 2003) and other presentation software tools. The internet is full of great examples including presentations with music, video snippets, and rotatable objects. In 2008 Adobe introduced Acrobat 9 Pro, which includes tools for embedding multimedia objects such as interactive 3D graphics, sound, and video. As an example, the paper by Goodman et al. (2009),

    Goodman et al., 2009, A role for self-gravity at multiple length
    scales in the process of star formation. Nature, 457, 63-66.

includes interactive 3D page (Figure 2). The release of the free-of-charge iBooks Author software by Apple in 2012 provided a *what you see is what you get* (WYSIWYG) type of editor for interactive ebooks in a proprietary file format similar to the open *electronic publication standard* (EPUB) ebook file format. The sister book to this book,

    Trauth, M.H. (2015) MATLAB Recipes for Earth Sciences - Fourth
    Edition. Springer, 427 p., Supplementary Electronic
    Material, Hardcover, ISBN: 978-3-662-46244-7.

was released as an interactive ebook, available from the Apple iBooks Store. This version of the book includes interactive 3D graphics, sound, and video, as well as an interactive review (or quiz) at the end of each chapter that tests the reader's understanding of its content.

Within this chapter, ►Sect. 12.2 demonstrates how to create and edit audio and video files, and then ►Sect. 12.3 introduces interactive 3D graphics objects. All of these are then included in multimedia presentations (►Sect. 12.4), in papers or books (►Sect. 12.5).

12.2 **Creating Audio and Video Files with MATLAB**

We will first create some audio files and then integrate them into multimedia publications. In theory, any audio material can be used including, for example, recordings of the sounds of nature. MATLAB contains many examples of audio material, such as the sound of a Chinese gong (contained in the file *gong.mat*) and the Halleluja Chorus from Handel's Messiah (contained in the file *handel.mat*) that can be used as examples. However, with MATLAB we are creating very simple files such as sine waves with noise exported as audio files and animated graphics exported as video files. To do this we first create a time axis t running from 1 to 30,000 in steps of one unit. We then generate a simple periodic signal $x1$ in the form of a sine wave with a period of 100 and an amplitude of 1, by typing

```
clear, clc
t = 0 : 30000;
x1 = sin(2*pi*t/100);
```

We can then convert the signal to sound using `sound`

```
sound(x1,44100)
```

where `44100` corresponds to the sampling rate of 44,100 cycles per second or 44,100 Hz (Hertz). Using this sampling rate, each of the data points in t corresponds to 1/44,100 s. The period of 100 therefore produces a tone with a frequency of 44,100 Hz/100 = 441 Hz. Natural data series, however, are more complex than a simple periodic signal. A slightly more complicated signal can be generated by superimposing several periodic components with different periods. As an example we compute such a signal $x2$ by adding three sine waves with periods of 300, 100 and 45 by typing

```
x2 = 0.2*sin(2*pi*t/300) + ...
     0.1*sin(2*pi*t/100) + ...
     0.3*sin(2*pi*t/45);
```

The corresponding amplitudes are 0.2, 0.1 and 0.3. Again, we can calculate the frequencies of the tones produced by

```
sound(x2,44100)
```

by dividing the sampling frequency of 44,100 Hz by the periods of the sine waves (300, 100 and 45), which yields 147, 441 and 980 Hz. In contrast to our synthetic time series, real data also contain various disturbances such as random noise. In order to reproduce the effects of noise, a random-number generator can be used to compute Gaussian noise with a mean of zero and a standard deviation of one. The seed of the algorithm should be set to zero using `rng(0)`, before then using `randn` to generate the noise. We add this noise to the original data, i.e., we generate a signal containing additive noise, by typing

```
x3 = x2 + 0.1*randn(size(x2));
```

and again convert the signal into sound:

```
sound(x3,44100)
```

We can create a *Waveform Audio File Format* (WAVE) file from the three signals using

```
audiowrite('aaudio_1.wav',x1,44100)
audiowrite('aaudio_2.wav',x2,44100)
audiowrite('aaudio_3.wav',x3,44100)
```

where `44100` again corresponds to the sampling rate. According to Wikipedia, WAVE (or WAV) is a Microsoft and IBM audio file format standard for storing an audio bitstream on PCs, mostly used on Windows systems for raw and typically uncompressed audio. The WAV format is compatible with other operating systems, such as macOS and Linux, and can be easily converted to other file formats. The function `audio-write` also supports other file formats, such as the MPEG-4 format. MPEG stands for the *ISO/IEC Moving Picture Experts Group*, i.e. the format takes the name of the group that developed it, as was also the case with the JPEG format (*Joint Photographic Experts Group*, see ▶Sect. 8.3). MPEG-4, which replaced the previous MPEG-1 and MPEG-2 standards, is a method for compressing audio and video data for the use in many different applications. MPEG-4 includes the *Advanced Audio Coding* (AAC) standard for lossy digital audio compression for audio files, typically achieving a better sound quality than the older MP3 standard contained in the MPEG-1 and MPEG-2 formats.

There are a number of software solutions available for editing audio files. We typically use such a software simply to cut audio files, in other words to shorten them by cutting at either one end, or at both ends. Software tools are supplied with the different operating systems specifically for this purpose, e.g. Apple's *QuickTime Player*, for computers running macOS. Others are available for free download, e.g. *VLC Media Player*, available for macOS, Windows and Linux. An advanced, very popular, and free software for editing audio files in macOS, Windows and Linux is *Audacity*. According to the Audacity webpage

```
http://www.audacityteam.org
```

it was developed by Dominic Mazzoni and Roger Dannenberg in the fall of 1999, at Carnegie Mellon University. It was released as open-source software through *Source-Forge.net* in May of 2000. Audacity can be used to record live audio, digitize recordings from other media, import and export various audio file formats, and analyze, edit and manipulate sound in many different ways. The commercial alternative to Audacity is *Adobe Audition CC*, included in the Adobe Creative Cloud software suite and first released in 2003:

```
http://www.adobe.com/products/audition.html
```

According to the product page on the internet, Adobe Audition is a professional audio workstation for mixing, finishing, and precession editing. Since the software is well integrated in the Adobe Creative Cloud software suite, it is a powerful tool for editing soundtracks during the video production workflow and audio finishing.

Next we will create some video files and integrate them into multimedia publications. In theory, any video material can be used including, for example, recorded laboratory and field experiments, screen video captures, or recorded computer animations. As an example, the sister book *MATLAB Recipes for Earth Sciences* (Trauth 2015) includes recorded screen activities to demonstrate the use of MATLAB. This material was created with the software *Any Screen Record Pro* for macOS, freely available through the Apple App Store. A popular free software solution for both macOS and Windows is the *Icecream Screen Recorder*

```
http://icecreamapps.com/Screen-Recorder/
```

The book *MATLAB Recipes for Earth Sciences* also contains numerous animated graphics, which are included as videos. As an example, ▶Sect. 6.4 of the book demonstrates the convolution of a single one within a series of zeros (a unit impulse series, marked with blue circles) with a running mean of length five, using the function `conv`. The convolution is performed by calculating a series of means (orange circles) for subsets of five data points from the original data series (orange rectangles) (◘ Fig. 12.1). In order to create this video we first create the input data series by typing

```
clear, clc
t = 1 : 50;
x = [zeros(24,1);ones(1,1);zeros(25,1)];
y = NaN(size(x));
```

We then define the size of the window, i.e. the length of the running mean.

```
w = 5;
```

Next we prepare the video file *amovie_1.avi* using `VideoWriter`. We define a frame rate of 1 frame per second and a video quality of 100%. We then check the parameter settings and open the file.

```
v = VideoWriter('amovie_1.avi');
v.FrameRate = 1;
v.Quality = 100;
v
open(v);
```

We can then run the animation and record the video by typing

```
j = 1;
for i = (w + 1)/2 : length(x)-(w-1)/2
    close all
    figure('Color',[1 1 1],...
            'Position',[300 300 1600 800])
    y(i) = mean(x(i - (w-1)/2 : i + (w-1)/2));
    y(length(x)-(w-1)/2 + 1:length(x)) = NaN;
    p3 = patch([i-(w-1)/2 i + (w-1)/2 ...
        i + (w-1)/2 i-(w-1)/2],...
            [-2 -2 2 2],'r',...
            'FaceColor',[0.8510 0.3255 0.0980],...
            'FaceAlpha',0.3); hold on
    s1 = stem(t,x);
    s2 = stem(t,y);
    s3 = stem(t(i - (w-1)/2 : i  +  (w-1)/2),...
        x(i - (w-1)/2 : i  +  (w-1)/2),'fill');
    s4 = stem(t(i),y(i),'fill');
    set(gca,'FontSize',20)
    set(s1,'MarkerSize',12)
    set(s2,'MarkerSize',12)
    set(s3,'MarkerSize',12,...
            'Color',[0    0.4471    0.7412]);
    set(s4,'MarkerSize',12,...
            'Color',[0.8510 0.3255 0.0980]);
    drawnow
    M = getframe(gcf);
    j = j + 1;
    writeVideo(v,M);
    hold off
end
```

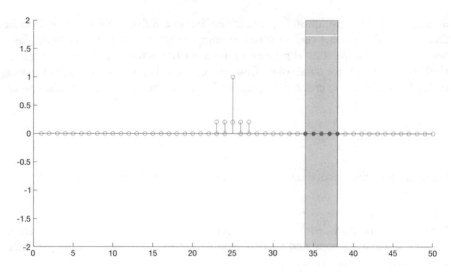

⬛ Fig. 12.1 Screenshot of the video example from Trauth (2015). The animation demonstrates the convolution of a single one within a series of zeros (a unit impulse series, marked with blue circles) with a running mean of length five. The convolution is performed by calculating a series of means (orange circles) for subsets of five data points from the original data series (orange rectangles)

Finally, we close the video file by typing

```
close(v)
```

The function `writeVideo` exports the animation in the Audio Video Interleave (AVI) format introduced by Microsoft in 1992 (⬛Fig. 12.1). We can view, edit and convert the video using the *QuickTime Player* (for computers running macOS), the *VLC Media Player* (available for macOS, Windows and Linux), or any other video player. More advanced video editors are available both for free download (e.g., *Lightworks* for macOS, Windows and Linux, *iMovie* for macOS) or for purchase (e.g., *Adobe Premiere Pro CC* for both macOS and Windows).

12.3 Creating Animated 3D Objects with MATLAB

Since the introduction of electronic devices with touch controls, interactive 3D graphics objects have become increasingly popular in multimedia electronic books (*ebooks*). The Simulink 3D Animation available from The MathWorks, Inc. provides the necessary tools to create and export 3D graphics objects for inclusion in documents such as multimedia ebooks, interactive webpages, and presentations (MathWorks 2016). A VRML file is a text file with the file extension.*wrl* for *world*, the accepted abbreviation for *Virtual Reality World*. Opening such a file in, for example, the MATLAB editor reveals that it contains vertices and edges for 3D polygons, together with parameters such as surface colors, textures, transparency, and so on. A VRML file can be viewed using a browser plugin (of which there are many different types available online), or using 3D software such as the open-source *MeshLab* software

```
http://www.meshlab.net
```

or the free *Blender* software available at

```
http://www.blender.org
```

These tools (as well as others) can be used to convert the VRML format into other 3D graphics object file formats such as the *Universal 3D or U3D* (file extension *.u3d*) and *COLLADA DAE* (file extension *.dae*) formats. The U3D format is the format required to place a 3D graphics object onto a PDF page with Adobe Acrobat, while the DAE format is used to include 3D graphics objects in Apple iBooks Author files, for use with Apple iPads and other iOS devices that have touch controls.

The digital terrain models created in previous sections can be used as examples of such interactive 3D objects. We use a modified script to import and display an SRTM data set from the previous section, as an example. The data are imported into the workspace using

```
clear

fid = fopen('S01E036.hgt','r');
SRTM = fread(fid,[1201,inf],'int16','b');
fclose(fid);
```

The matrix first needs to be transposed and flipped vertically.

```
SRTM = SRTM';  SRTM = flipud(SRTM);
```

The SRTM data contain numerous gaps that might cause spurious effects during statistical analysis or when displaying the digital elevation model as a 3D graphics. We can use the method described in the previous section to eliminate the gaps. In this case, however, we have to search the data set for additional possible NaNs, even if we have already searched for NaNs and replaced them with the means of the surrounding pixels, because the VRML clients cannot handle the character string *NaN* and therefore produce an error message when importing the file into software such as MeshLab. Since the resulting plot will be for display purposes only, we can also use a simplified method that involves replacing the gaps marked by −32768 with the mean value of the entire DEM data set.

```
SRTM(find(SRTM == -32768)) = mean(SRTM(:));
```

A coordinate system can then be defined using the information that the coordinates of the lower-left corner are s01e036. The resolution is 3 arc seconds, corresponding to 1/1,200 of a degree.

```
[LON,LAT] = meshgrid(36:1/1200:37,-1:1/1200:0);
```

We also need to reduce the size of the array by a factor of 10 because most VRML clients limit the number of vertices in a VRML file.

```
LON = LON(1:10:end,1:10:end);
LAT = LAT(1:10:end,1:10:end);
SRTM = SRTM(1:10:end,1:10:end);
```

You can run an experiment with a larger number of vertices and see whether it works with your VRML software. We also eliminate the edges of the data set, as these may cause problems when writing the VRML files.

```
LON = LON(2:end-1,2:end-1);
LAT = LAT(2:end-1,2:end-1);
SRTM = SRTM(2:end-1,2:end-1);
```

3D graphics can now be generated from the elevation data using the function trimesh following Delaunay triangulation of the data set using delaunay.

```
tri = delaunay(LON,LAT);
trimesh(tri,LON,LAT,SRTM)
axis([35.5 37.5 -1.5 0.5 -500 4500]), axis off
```

Next, we can determine the size of the array SRTM, which is found to be 119-by-119, and then reshape it to an n-by-1 array, where $n = $ xdim*ydim $= 14161$.

```
[xdim ydim] = size(SRTM);
SRTM = SRTM(:);
```

We then determine the range of the z-values and define the spacing in x and y directions.

```
zrange = range(SRTM);
xspace = 10;
yspace = 10;
```

We use the colormap demcmap to display the SRTM data set.

```
cmap = demcmap(SRTM,256);
```

Alternatively, we can use any other colormap, even one that we have created ourselves, to display the digital terrain model. We then normalize the elevation data to the [1,length(cmap)] range.

```
cmap = cmap(round((SRTM-min(SRTM)) ...
    .*(size(cmap,1)-1)./zrange)+1,:);
```

The Simulink 3D Animation User's Guide (MathWorks 2016) contains a great introduction in its chapter entitled *Interact with Virtual Reality Worlds*, which explains the following steps in greater detail than provided here. We first need to make sure that all virtual reality worlds are closed and deleted.

```
out = vrwho;
for i=1:length(out)
    while (get(out(i),'opencount') ~=0)
        close(out(i));
    end
    delete(out(i));
end
```

We then create a new virtual reality world `myworld` using `vrworld`.

```
myworld = vrworld('');
open(myworld)
```

Our world then needs a shape and a grid. We create the shape `newShape` with the name `Landscape` and the grid `newGrid` using `vrnode`. The function `vrnode` creates a handle for either an existing node or a new node.

```
shapeName = ['Landscape'];
newShape = vrnode(myworld,shapeName,'Shape');
newGrid = vrnode(newShape,'geometry','DEM','ElevationGrid');
```

We can access the contents of the variable `newShape` from the `vrworld` class in much the same way that we access the contents of structure arrays, e.g., by typing

```
getfield(newShape.geometry)
```

which lists a number of the defining parameters of `newShape` such as, for example, the color. The nodes of the world can be accessed using one of these two commands:

```
nodes(myworld)
mynodes = get(myworld,'Nodes')
```

We can access the fields of the node `Landscape` using one of these two commands:

```
fields(myworld.Landscape)
fields(mynodes.DEM)
```

and the `DEM` by using

```
fields(myworld(1))
fields(mynodes(2))
```

In each case both options yield a detailed output of all the parameter settings of the nodes. We can also list the fields in `newShape` and `newGrid` using

```
fields(newShape)
fields(newGrid)
```

and modify the values of the various parameters of `newShape` and `newGrid`. As an example we can reduce the heights to 20% of the original SRTM elevations in order to change the proportions of the 3D graphics.

```
setfield(newGrid, ...
       'xDimension',xdim,...
       'zDimension',ydim,...
       'xSpacing',xspace,...
       'zSpacing',yspace,...
       'height',0.2*SRTM);
```

We then use the same colormap `cmap` that was created previously (before using `demcmap`), based on elevation

```
GridColor = vrnode(newGrid,...
        'color','TerrainColor',...
        'Color');
GridColor.color = cmap;
getfield(newGrid,'color')
```

and save the world `myworld` created with Simulink 3D Animation to a new VRML file *srtm_2.wrl*.

```
save(myworld,'srtm.wrl')
close(myworld)
delete(myworld)
```

We then close and delete the world `myworld`. Importing the resulting VRML file into a VRML client such as MeshLab or Blender reveals that the terrain model does indeed have colors, as defined by `demcmap`. We can use the file in the VRML format to create an interactive document. However, the software to create such interactive documents, such as Adobe Acrobat or Apple's iBooks Author, may require different file formats. In this case the VRML format can easily be converted into the required format using MeshLab or Blender software.

12.4   Assembling Multimedia Presentations

Multimedia objects, including audio or video material, can enrich an oral presentation tremendously. Our example presentation will be created with the free *OpenOffice Impress* software, but working with Microsoft PowerPoint and Apple Keynote is very similar. We launch OpenOffice and choose *Presentation* from the Start Center. The OpenOffice *Presentation Wizard* pops up and asks us to choose between an empty presentation, a presentation from a template, and an existing presentation. We will ignore the template gallery and design our own simple template, so we choose *Empty presentation*. We then select *<Original>* for the background for our presentation and click on *Next ≫*. On the next page we choose *No Effect* for the effect and *Default* for the presentation type, and then click on *Create*.

The software creates a new document entitled *Untitled 1*, which we can save as *multimediapresentation_vs1.odp*. We choose the empty slide from the *Layouts* panel on the right. We can add a second slide by choosing *Slide* from the *Insert* menu. We can now import the audio file by choosing *Movie and Sound* from the *Insert* menu and then selecting the file *aaudio_1.wav* from the hard drive. Drag and drop also works, i.e. we can select the file by clicking on the file icon, drag it over the empty slide, and then drop it there. The same methods works with the movie stored in *amovie_1.mov*, which we can either be inserted by using *Movie and Sound* from the *Insert* menu, or by dragging and dropping the file onto the slide. We can move and scale the objects on the slide, add text and other objects to the slide, and then examine the result using *Slide Show* from the *Slide Show* menu. In addition to.*wav* and.*mov* file formats, a number of other formats for audio files and movies can be used with OpenOffice Impress.

Importing movies and sounds with *Microsoft PowerPoint* yields slightly more professional-looking results. We again create a new presentation by choosing *New Presentation* from the *File* menu, save it as *multimediapresentation_vs1.pptx*, and then remove unnecessary text boxes before importing the examples by either selecting *Audio* or *Video* from the *Insert* menu to import the examples, or dragging the audio file from *aaudio_1.wav* and the movie from *amovie_1.mov* onto the empty slide and dropping them there. We can again move and scale the multimedia objects and add text to the slides. We choose *Play from Start* from the *Slide Show* menu to examine the result.

In *Apple Keynote* we create a new presentation by choosing *New ...* from the File menu, and then we select the *White* presentation. After removing unnecessary text boxes on the first slide and creating a second empty slide, we can import the audio file and movie using *Choose ...* from the *Insert* menu. The multimedia files can be in different formats, such as *.wav* or *.mp3* formats for the audio files, and *.mov* or *.avi* for the movies. Selecting the sound object on the slide and clicking on the *Audio* tab on the right, the *Audio* panel appears and we can change the volume, trim the length, and control the looping. Similarly, by clicking on the movie, and then on the *Movie* tab, we can do the same with the movie. Choosing *Play Slideshow* from the *Play* menu allows us to examine the result.

It is unfortunately not yet possible to include animated 3D objects in Powerpoint and Keynote presentations. Such objects can, however, be embedded into presentations using the Adobe Creative Cloud software suite. Adobe InDesign allows various types of objects, including movie and sound objects, to be placed into position and then a PDF file created in landscape view. We create a new document with dimensions of 1,920 × 1,080 pixels, add a second page, and then save the document as *multimediapresentation_vs1.indd*. We can then place the movie file *amovie_1.mov* and the audio file *aaudio_1.mp3* onto the pages of the document. We export the document as *multimediapresentation_vs1.pdf*, which we can then open with the Acrobat Reader and present in full screen mode.

Adobe InDesign does not allow us to insert animated 3D objects but we can use the file we just created, or create a new one, and use Adobe Acrobat instead to add multimedia objects. After opening *multimediapresentation_vs1.pdf* we use the *Rich Media* toolbar of Adobe Acrobat, which allows us to add animated 3D objects, push buttons, audio and video files, and *Shockwave Flash* (SWF) objects. We add a third page to the document and import the animated 3D object from the file *srtm.u3d*, which is in the *Universal 3D.u3d* format. This file has been created by converting the VRML file *srtm.wrl* using the *MeshLab* or *Blender* software (▶Sect. 12.3). Right-clicking on the imported object allows us to modify a number of properties, such as the background color (e.g. from dark gray to white) and the default view. Saving the document and then opening it with Acrobat Reader in the full screen mode allows us to see that we can actually use the movie, the sound, and the 3D object interactively.

The best option at the moment is the macOS-only iBooks Author software, which was actually developed for the creation of interactive ebooks. After launching the software we select the *Blank* template from the *Template Chooser*, add two more pages using the *Add Pages* button, and save the file as *multimediapresentation_vs1.iba*. We will go into the detailed use of this software in the next section, but for now we will restrict ourselves to briefly discussing the embedding of interactive objects. For this we use *Widgets*, which can be accessed via the button of the same name in the toolbar. Clicking on Widgets allows us to choose between *Gallery, Media, Review, Key-*

Signal of a single
sine wave

Signal of three sine
waves

Signal of sine waves
with noise

Fig. 12.2   Screenshot of Page 2 of a multimedia presentation created with iBooks Author. The page, which contains three audio examples, is similar to Page 286 of the interactive ebook of MATLAB Recipes for Earth Sciences, 4th Edition for iOS devices and computers running macOS (Trauth 2015)

note, *Interactive Image, 3D, Scrolling Sidebar, Pop-Over* and *HTML*. The *HTML* option in particular permits considerable extensions to be made to the software, but that is beyond the scope of the book.

Selecting *Media* from the *Widget* menu places onto the page an empty movie object (including header and footer), into which we can drag and drop the file *amovie_1. mov*. We can change the blind text of the header and footer, the position and size of the object, and many other things. We use the same widget to add sound from the file *aaudio_1.mp3*. Once again, we can change the blind text of the header and footer, the position and size of the object, and so forth. Having finished editing the file we use *Export …* from the *File* menu to create a *.ibooks* file. This file can be viewed with the *Apple iBooks* software, which is used to manage and read all ebooks (▶ http://www. apple.com/de/ibooks). After launching the software we can open the file *multimediapresentation_vs1.ibooks* and display it in the full-screen mode. This is therefore a good way to include sound, movie and animated 3D objects in a presentation (☐ Fig. 12.2).

12.5 Assembling Multimedia Papers and Books

With its fourth edition the sister book to this book, *MATLAB Recipes for Earth Sciences* (Trauth 2015), made its first appearance as an interactive ebook in which the reader is able to interact with the book. It comes in two different formats, with the first being for Apple iPads and Macs (in the form of a .ibooks file created with Apple's iBooks Author software) and the second being for other platforms (as a PDF file that includes relative links to interactive objects stored on the reader's hard drive). The ebook has been designed to be read in landscape mode; it includes movies, galleries, audios and interactive 3D displays, as well as reviews at the end of each chapter.

The movies demonstrate the use of graphical user interface tools; they also help to explore the effect that changing input parameters has on the output of a function, and to visualize mathematical operations. Galleries are generally used to present a series of associated graphics, such as those displaying the various parameters describing the dispersion and shape of a distribution. Audios are used to provide striking representation of signals and noise, as well as the effect of filters. Interactive 3D displays allow interactive rotation of digital terrain models and other three-dimensional objects. As mentioned previously, each chapter ends with a review (or quiz) to test the reader's understanding of its content. These interactive media (or widgets) can be explored, observed, or listened to on Macs using a multi-touch trackpad or a mouse, or on iPads using your finger(s) by tapping the *Play* button (to watch movies or to listen to audios), by tapping on the arrows or swiping left or right (to navigate through a gallery), or by tapping on and dragging a 3D object to rotate it. Movies, galleries and 3D objects can be zoomed to full screen by using the full-screen toggle in the lower right corner (with movies), or by simply tapping on the image (with galleries and 3D objects).

The interactive ebook for other platforms comes as a PDF file, which can be accessed with the free Adobe Reader software. The reader receives a complete package of files that includes not only the PDF version of the ebook, but also a series of directories containing the interactive objects, the recipes with all the MATLAB commands featured in the book, and the example data. The interactive objects are stored separate from the PDF file and can be accessed from the *MRES4-Index.html* file, which can be viewed in a web browser. The interactive PDF version of the ebook contains all of the interactive objects included in the ebook for iPads and Macs, except for the reviews. It was planned as an ebook with embedded interactive objects such as movies, galleries, audios and reviews, but during the production process it soon became clear that it was not yet possible to offer such a product that would be compatible with all platforms.

Creating interactive books with the Adobe Creative Cloud suite is therefore not included as a subject in this edition. An interactive ebook with audio, video and interactive 3D objects is instead created using Apple's iBooks Author, which is far more suitable for this task. After launching the software we select the *Blank* template from the *Template Chooser* and save the file as *multimediabook_vs1.iba*. Alternatively, we can choose any other template from the Template Chooser or use one of the templates available on the Internet. Templates have predesigned page layouts, which in the case of the white template is very Spartan. The template contains the first page of Chapter 1 and the first page of Section 1. We can drag and drop a photo onto the first page of the Chapter (e.g. *multimediabook_photo1.jpg*), scale it to fit the page, and use *Send to Back* from the *Arrange* menu to move it to the lowest level of the page. We can then change the title of our first chapter to *Title of Chapter 1*, and change the font color to white. If we want to we can write a few sentences to describe the contents of the chapter in the textbox beneath the title, and again change the font color to white. We then add text to the two-column text box of the Section 1 page. We can use a blindtext editor to create 1,000 words of *lorem ipsum* text (▶ http://www.blindtextgenerator.com). Since our *lorem ipsum* text is quite long the software automatically adds a a second page to Section 1, with more textboxes.

After having inserted our text, we now include illustrations. Selecting *Gallery* from the *Widget* menu places an empty photo object on the page (including header and footer) into which we can drag and drop the file *multimediabook_photo2.jpg*. We use the *Inspector* from the toolbar to change the layout of the widget to *Bottom*, which removes the header above the photo and only displays a figure caption below the photo. We can add text to the figure caption, such as *Fig. 1 Sandstone hoodoo …* and then save the file. The *Gallery* widget allows us to add more photos and accompanying captions, thus creating a gallery on the page. We can also include multimedia such as sound, movie and interactive 3D objects, as demonstrated in the previous section.

Section 1
Title of Section 1

Lorem ipsum dolor sit amet, consectetuer adipiscing elit. Aenean commodo ligula eget dolor. Aenean massa. Cum sociis natoque penatibus et magnis dis parturient montes, nascetur ridiculus mus. Donec quam felis, ultricies nec, pellentesque eu, pretium quis, sem. Nulla consequat massa quis enim. Donec pede justo, fringilla vel, aliquet nec, vulputate eget, arcu. In enim justo, rhoncus ut, imperdiet a, venenatis vitae, justo. Nullam dictum felis eu pede mollis pretium. Integer tincidunt. Cras dapibus. Vivamus elementum semper nisi. Aenean vulputate eleifend tellus. Aenean leo ligula, porttitor eu, consequat vitae, eleifend ac, enim. Aliquam lorem ante, dapibus in, viverra quis, feugiat a, tellus. Phasellus viverra nulla ut metus varius laoreet. Quisque rutrum. Aenean imperdiet. Etiam ultricies nisi vel augue. Curabitur ullamcorper ultricies nisi. Nam eget dui. Etiam rhoncus. Maecenas tempus, tellus eget condimentum rhoncus, sem quam semper libero, sit amet adipiscing sem neque sed ipsum. Nam quam nunc, blandit vel, luctus pulvinar, hendrerit id, lorem. Maecenas nec odio et ante tincidunt tempus. Donec vitae sapien ut libero venenatis faucibus. Nullam quis ante. Etiam sit amet orci eget eros faucibus tincidunt. Duis leo. Sed fringilla mauris sit amet nibh. Donec sodales sagittis magna. Sed consequat, leo eget bibendum sodales, augue velit cursus nunc, quis gravida magna mi a libero. Fusce vulputate eleifend sapien. Vestibulum purus quam, scelerisque ut, mollis sed, nonummy id, metus. Nullam

Fig. 1 Sandstone hoodoo in the Baragoi river valley, East of the Suguta Valley, Kenya.

◼ **Fig. 12.3**   Screenshot of a page of the ebook created with iBooks Author, showing a photo gallery within a two-column text box

In a very similar way to other software tools for creating books, iBooks Author allows you to design a book cover and to create a table of contents, a glossary, and many other things. A movie can also be included after the title page with *Intro Media*, as a very attractive way of introducing the subject matter of the book. Having finished editing the document we use *Export …* from the *File* menu to create a *.ibooks* file. This file can be viewed with the *Apple iBooks* software, which is used to manage and read all ebooks (▶http://www.apple.com/de/ibooks). After launching the software we can open the file *multimediapresentation_vs1.ibooks* and display it in full-screen mode (◼Fig. 12.3). This method could also be used to include sound, movie, and animated 3D objects in a presentation. Apple's workflow is directed towards publishing the book through Apple's iBooks store, but you can also publish the *.books* file yourself.

Recommended Reading

Goodman AA, Rosolowsky EW, Borkin MA, Foster JB, Halle M, Kauffmann J, Pineda JE, (2009) A role for self-gravity at multiple length scales in the process of star formation. Nature 457 (7225):63–66
MathWorks (2016) Simulink 3D Animation—User's Guide. The MathWorks, Natick
Trauth MH (2015) MATLAB® Recipes for Earth Sciences–4th Edition. Springer, Berlin

Printed in the United States
By Bookmasters